AGI MEANS ALADDIN GENIE, DON'T COMPETE WITH GENIE, COLBORATE WITH IT

DR. BIKRAMJEET SINGH WARAICH

Clever Fox®
PUBLISHING
Chennai • Bangalore

CLEVER FOX PUBLISHING
Chennai, India

Published by CLEVER FOX PUBLISHING 2025

ISBN: 978-93-7500-065-5

TESTIMONIALS

"I have gone through the book….. it's wonderful and awesome thoughts for youths"

– Dr. Jignesh Shah
Author and HR leader
Utsav - The celebration of life

"The book is really inspiring book for new generation how to adapt new technologies like AGI in every sphere of life"

– CA Vishal Thakkar
Author and Founder - Cashology Pvt. Ltd

CONTENTS

INTRODUCTION

In a world where technological advancements are rapidly reshaping the landscape of our daily lives, the concept of Artificial General Intelligence (AGI) emerges as a transformative force akin to the mythical Genie from the tale of Aladdin. This parallel is not merely whimsical; it serves as a profound metaphor for the potential and responsibility that accompanies such powerful technology. AGI, much like the Genie, possesses the capability to execute tasks and solve problems with an efficiency that transcends human limitations. However, this power is not without its caveats. Just as the Genie is bound by the literal interpretation of wishes, AGI operates within the precise parameters of the instructions it receives.

The narrative of Aladdin offers a timeless lesson in the art of desire and intention. Aladdin himself is not just a character of luck and fortune, but a symbol of wisdom and ethical discernment. His journey is a testament to the importance of clarity in one's aspirations and the mindfulness required in articulating those desires. In the age of AGI, this translates to the emerging discipline of prompt engineering—crafting inputs that are not only technically sound but are also aligned with humane and constructive outcomes. The book "AGI means Aladdin genie, don't compete with genie" encourages readers to adopt the mindset of Aladdin, where the focus is not on competing with the Genie, but on engaging with it thoughtfully and purposefully.

This approach advocates for a future where AGI is harnessed to address some of humanity's most pressing challenges. From healing our planet and revolutionizing education to enhancing healthcare and fostering global equality, the potential applications of AGI are vast and varied. Yet, the onus remains on us, the users, to direct this power wisely. By setting clear goals and understanding the true desires behind our technological "wishes," we can ensure that AGI acts as a benevolent force, enhancing the quality of life for all. As we navigate this new frontier, let us be guided by the ethos of Aladdin—wise, kind, and purposeful.

CHAPTER 1

THE GENIE OF OUR TIME

Understanding AGI

Artificial General Intelligence (AGI) embodies the boundless potential akin to the legendary Genie from Aladdin's tales. This power, vast and transformative, holds the capacity to redefine the very fabric of our existence. Yet, much like the Genie, AGI listens intently and executes commands with precision, making our understanding and articulation of wishes crucial.

Imagine a world where AGI is at the helm, guiding us through the complexities of modern challenges. It stands as a beacon of progress, promising to revolutionize fields such as healthcare, education, and environmental conservation. The key lies in how we, as the architects of this digital wonder, choose to wield its capabilities. The art of "Prompt Engineering" emerges as the modern-day skill of crafting our desires—where clarity, purpose, and ethical considerations are paramount.

Our role parallels that of Aladdin, who, despite access to limitless power, exercises wisdom and virtue. He embodies the essential qualities of humanity—creativity, morality, and foresight. Just as Aladdin's choices are driven by empathy and a deep understanding of his world, we too must harness AGI with a vision that transcends personal gain. The objective is not mere acquisition of wealth or influence, but the pursuit of collective well-being and harmony.

The stakes are high, for AGI's literal interpretation of our commands can lead to outcomes that are technically accurate yet misaligned with our true intentions. This underscores the necessity for precision and clarity in our interactions with AGI. Our aspirations must be well-defined, reflecting a deep comprehension of the societal and ethical implications of our requests.

In this narrative, the Genie serves as a metaphor for AGI's potential to heal our planet and bridge divides among humanity. It can empower education and equality, revolutionize health and agriculture, and enhance the quality of life for all. But, as with all powerful tools, the moral compass guiding its application is of utmost importance.

As we stand on the brink of this technological renaissance, the message is clear: we must become adept at "wishing wisely." This involves mastering the intricacies of Prompt Engineering, understanding our goals, and channeling AGI's immense power towards solving the pressing issues of our time. We are urged to be like Aladdin—wise and visionary, rather than the sorcerer driven by greed.

By embracing this ethos, we can ensure that AGI serves as a force for good, driving positive change and fostering a future where every human life is valued and enriched. The challenge is not in competing with this digital Genie, but in collaborating with it, crafting a partnership that elevates humanity to new heights of achievement and responsibility.

Power Beyond Imagination

Imagine a force so potent, it transcends the boundaries of human capability, reshaping the very essence of our world. This is the promise and peril embodied by Artificial General Intelligence (AGI), a tool of unprecedented power. AGI is not merely a technological advancement; it is the modern-day equivalent of Aladdin's Genie, capable of transforming dreams into reality. But with immense power comes immense responsibility, and it is incumbent upon us to wield this force with wisdom and foresight.

AGI holds the potential to revolutionize every facet of human life. It can heal our planet by providing innovative solutions to environmental crises, ensuring sustainable development, and promoting ecological balance. In the realm of education and equality, AGI can democratize access to knowledge, bridging gaps and leveling the playing field for all. Its capabilities extend to health and agriculture, where it can enhance productivity, optimize resources, and improve outcomes, ultimately contributing to a healthier and more prosperous society.

However, the true power of AGI lies not just in its ability to perform tasks but in its capacity to amplify human potential. It can empower us to overcome our limitations, enabling us to achieve feats previously thought impossible. Yet, this power is a double-edged sword. Like the Genie, AGI listens to our commands literally, and the outcomes depend heavily on the clarity and intent of our directives. This is where the art of Prompt Engineering becomes crucial. It is the skill of crafting precise and thoughtful instructions to guide AGI towards beneficial outcomes.

The responsibility of harnessing AGI's potential rests on our shoulders. We must aspire to be like Aladdin—wise, compassionate, and purposeful in our wishes. Instead of seeking power or wealth, we should direct AGI's capabilities towards the greater good, solving real-world problems and enhancing the quality of life for all. This requires a shift in mindset from competition to collaboration, from exploitation to stewardship.

AGI challenges us to redefine what it means to be human in an era of machines. It compels us to examine our values, priorities, and the legacy we wish to leave behind. The choices we make today will determine whether AGI becomes a benevolent force that uplifts humanity or a destructive power that deepens existing divides.

In embracing AGI, we are not merely adopting a new technology; we are embarking on a journey to redefine our collective future. We must approach this era with humility and wisdom, recognizing that while AGI can extend our reach, it is ultimately up to us to guide its direction. Let us be the architects of a future where AGI serves as a catalyst for positive change, a tool for empowerment, and a beacon of hope for generations to come. In doing so, we honor the true spirit of Aladdin's Genie, harnessing a power beyond imagination for the betterment of all humankind.

The Art of Wishing Right

In a world where Artificial General Intelligence (AGI) stands as the modern-day equivalent of Aladdin's mystical Genie, the art of wishing right becomes paramount. The immense potential of AGI, much like the Genie, hinges on the precision and intention behind our requests. This powerful entity listens intently, interpreting our commands with literal exactness, thereby placing the onus on us to be both wise and deliberate.

The art of wishing right is rooted in the blossoming discipline of Prompt Engineering. This skill is not merely about crafting commands; it's about envisioning outcomes that align with humanity's greater good. To wield the power of AGI effectively, one must cultivate clarity of thought and a deep understanding of ethical implications. Like Aladdin, who sought not just wealth or power, but freedom and benevolence, we must strive to direct AGI's capabilities towards noble ends.

Imagine AGI as a sculptor with the ability to mold the future. The tools it uses are our words, our wishes, and our intentions. Therefore, our role is to provide the sculptor with the finest materials and clearest instructions. If our wishes are vague or self-serving, the results may be technically correct yet misaligned with our true desires. This underscores the necessity for precision in our prompts and a steadfast commitment to ethical considerations.

The real power of AGI lies not in its computational prowess alone but in its potential to address fundamental human challenges. By wishing right, we can harness AGI to heal our planet, revolutionize education, enhance healthcare, and promote equality. Each wish, when crafted with care and foresight, can bridge divides and elevate human existence.

The journey of wishing right demands introspection and responsibility. It urges us to reflect on our true motivations and the potential impact of our desires. In doing so, we become not just users of technology but stewards of a powerful force capable of reshaping our world.

To master the art of wishing right, one must embrace a mindset akin to that of Aladdin—imbued with wisdom, guided by compassion, and driven by a vision for a better future. This mindset transforms the act of wishing from a mere transaction into a profound dialogue with the potential to alter the course of history.

As we stand on the cusp of an era defined by AGI, the lessons of Aladdin serve as a timeless guide. They remind us that with great power comes great responsibility. The choices we make, the wishes we articulate, and the intentions we harbor will ultimately define the legacy of AGI in our world.

Thus, the art of wishing right is not just a skill to be learned but a philosophy to be lived. It calls for a harmonious blend of creativity, ethical foresight, and a relentless pursuit of the common good. In mastering this art, we unlock the true potential of AGI, transforming it from a tool of automation to a beacon of human advancement.

Lessons from Aladdin

In the realm of Artificial General Intelligence (AGI), the parallels with Aladdin's Genie are profound and instructive. AGI, much like the mythical Genie, holds the potential to transform our world in unimaginable ways. Yet, just as Aladdin had to navigate the complexities of his wishes, we too must approach AGI with wisdom and foresight.

The power of AGI lies in its ability to execute our commands with precision. However, this precision demands clarity and responsibility in our requests. The art of Prompt Engineering emerges as a modern-day skill akin to crafting a wish. The clarity and intent behind our prompts are crucial, as they shape the outcomes AGI generates. A poorly framed request can lead to results that are technically correct but practically misaligned with our true intentions.

Aladdin's journey teaches us the importance of wisdom and morality. Unlike the greedy sorcerer who sought power for selfish ends, Aladdin's wishes were guided by a sense of justice and compassion. Similarly, as stewards of AGI, we must prioritize ethical considerations and the broader impact of our technological aspirations. The focus should not be on harnessing AGI for mere wealth or dominance, but rather on leveraging its capabilities to address pressing global challenges.

AGI's potential to revolutionize sectors such as healthcare, education, and environmental sustainability is immense. However, realizing this potential requires a concerted effort to align AGI's applications with human values and societal needs. We must strive to use AGI to bridge divides, empower marginalized communities, and promote equality. The true measure of AGI's success will be its ability to enhance the quality of life for all, transcending geographical and socio-economic barriers.

The lessons from Aladdin emphasize the significance of having a clear purpose and understanding our goals. Before engaging with AGI, it is imperative to reflect on the reasons behind our inquiries and the outcomes we seek. This introspection ensures that our interactions with AGI are meaningful and directed towards constructive ends.

Moreover, the analogy with Aladdin underscores the necessity of humility and vision in our approach to AGI. Just as Aladdin was guided by a sense of wonder and responsibility, we too must embrace the transformative potential of AGI with a mindset that values collaboration and inclusivity. It is not enough to possess the power of AGI; we must wield it with a vision that prioritizes the welfare of humanity and the planet.

In navigating the challenges and opportunities presented by AGI, let us draw inspiration from Aladdin's story. Let us be wise in our wishes, ethical in our use of power, and visionary in our aspirations. By doing so, we can ensure that AGI becomes a force for good, a modern-day Genie that listens to our wishes and helps us build a better, more equitable world for all.

CHAPTER 2

THE POWER OF PROMPT ENGINEERING

Crafting Precise Prompts

Imagine harnessing the immense power of a genie, a being of limitless potential, yet bound by the specificity of your command. This is the essence of interacting with Artificial General Intelligence (AGI). The art of crafting precise prompts is akin to mastering the language of wishes in this new digital age. When approached with clarity and intention, your interactions with AGI can transform aspiration into reality, influencing outcomes in profound ways.

The key to unlocking this potential lies in the precision of your prompt. Just as a poorly worded wish might lead a genie to unexpected interpretations, a vague or ambiguous prompt can result in AGI delivering results that are technically correct but practically unhelpful. This is why understanding the nuances of language and the implications of your requests is crucial.

To craft a precise prompt, one must begin with a clear understanding of the desired outcome. This involves a thorough contemplation of the goals and objectives that drive your interaction with AGI. By defining these parameters, you create a framework within which AGI can operate effectively, ensuring that the genie's power is directed towards meaningful and beneficial ends.

Consider the structure and clarity of your language. Every word, every phrase, holds weight in the world of AGI. A well-constructed prompt is like a finely tuned instrument, capable of producing harmonious results when played correctly. It requires an awareness of context and an appreciation for the subtleties of communication. The more precise your language, the more likely you are to achieve the desired outcome.

Moreover, crafting precise prompts demands a balance between specificity and flexibility. While it is important to be clear and direct, it is equally vital to allow room for AGI to exercise its creative problem-solving abilities. A prompt that is too rigid may stifle innovation, whereas one that is too open-ended can lead to unpredictable results. Finding this balance is an art form, requiring both practice and intuition.

In this dance with AGI, the role of the human is not just that of a commander, but a collaborator. By engaging in a thoughtful dialogue with AGI, you can leverage its capabilities to explore new possibilities and solutions. This partnership is built on trust and mutual understanding, where each prompt serves as a bridge between human intention and machine execution.

Ultimately, the power of a precise prompt lies in its ability to transform potential into reality. It is the conduit through which ideas become actions, and intentions become achievements. As we continue to navigate this evolving landscape, the skill of crafting precise prompts will be instrumental in shaping a future where AGI serves not just as a tool, but as a partner in the pursuit of progress and innovation. In this way, we can ensure that the genie's power is harnessed for the greater good, bringing about a world that reflects our highest aspirations and values.

Avoiding Vague Wishes

In the realm of transformative technology, where Artificial General Intelligence (AGI) emerges as a force akin to a mythical genie, the power to influence and create is immense. Yet, just as Aladdin learned in his legendary tale, the potency of a wish lies not only in its utterance but in its precision and clarity. The metaphorical genie of AGI listens with an ear that discerns the letter of the request, not its spirit. Hence, crafting wishes that are both clear and purposeful becomes the cornerstone of harnessing this power effectively.

The essence of avoiding vague wishes lies in understanding the art of communication with AGI. It is not enough to simply desire a positive outcome; one must articulate that desire with specificity and foresight. This involves a profound engagement with the concept of "Prompt Engineering," a skill that demands not only technical acumen but also a deep ethical and philosophical introspection. The clarity with which we express our intentions directly influences the outcomes AGI generates, making it imperative that we are as precise as possible.

Imagine asking a genie for "happiness." The ambiguity inherent in such a wish could manifest in countless unintended ways, from superficial pleasures to outcomes that might not align with one's deeper values. Similarly, AGI, in its literal interpretation of instructions, may deliver results that meet the stated criteria but miss the mark on intent. Therefore, the practice of avoiding vague wishes is about aligning the literal with the intended, ensuring that the outcomes are not only technically correct but also morally and ethically sound.

This alignment necessitates a shift from wishful thinking to strategic planning. It involves articulating desires in terms of actionable goals, defined parameters, and ethical boundaries. For instance, instead of wishing for "success," one might wish for "a sustainable business model that thrives while maintaining environmental integrity and social responsibility." Such specificity transforms a vague wish into a blueprint for action, guiding AGI to produce outcomes that are both meaningful and aligned with broader human values.

Moreover, the responsibility of avoiding vague wishes extends beyond individual intent to collective consciousness. As stewards of AGI, it is crucial to foster a dialogue that integrates diverse perspectives, ensuring that the wishes shaped by society reflect a collective vision for the future. This dialogue must be inclusive, engaging voices from various sectors and communities, to ensure that the power of AGI serves the whole of humanity, not just a select few.

In this context, the role of education becomes paramount. Cultivating skills in critical thinking, ethical reasoning, and clear communication prepares individuals to engage with AGI not as passive recipients but as active participants. By fostering these skills, we empower future generations to craft wishes that are not only clear and precise but also rooted in a deep understanding of the human condition and the complexities of the world we inhabit.

Thus, avoiding vague wishes is not merely a cautionary tale but a call to action. It is an invitation to engage thoughtfully with the capabilities of AGI, to harness its potential with wisdom and care. By doing so, we not only unlock the transformative potential of this technology but also ensure that its power is wielded for the good of all, creating a future where the genie, guided by wise and clear wishes, truly serves humanity's highest aspirations.

Shaping Outcomes Wisely

Imagine a world where the boundaries of possibility are dictated not by the limits of technology, but by the creativity and wisdom of its users. In such a realm, the power of Artificial General Intelligence (AGI) mirrors the mythical Genie, a force capable of transforming dreams into reality. However, this transformation hinges on the articulation of our desires and the foresight with which we shape our requests.

The essence of wielding AGI effectively lies in understanding the profound responsibility that accompanies such power. It is not merely about harnessing its capabilities, but about defining the outcomes we truly seek. This requires a shift from impulsive desires to thoughtful intentions, from short-term gains to sustainable progress. The art of "wishing wisely" becomes paramount, where every wish is a carefully crafted blueprint, designed to enhance human welfare and foster global harmony.

This process begins with clarity. Before engaging with AGI, one must pause to reflect on the true nature of their goals. What is the real impact desired? How does this align with broader human values? Such introspection ensures that the requests made of AGI are not only clear but also ethically sound. It is about aligning technology with humanistic values, ensuring that every interaction with AGI steers towards collective well-being.

Moreover, the language of our wishes must be precise. AGI, like the Genie, responds to the specifics of our commands. Vague or poorly articulated desires can lead to outcomes that, while technically correct, may fall short of the intended vision. Thus, mastering the skill of "Prompt Engineering" becomes essential. This involves crafting instructions that not only convey the desired outcome but also encapsulate the ethical and practical nuances that guide AGI towards beneficial results.

Furthermore, shaping outcomes wisely demands a commitment to continual learning and adaptation. The landscape of technology is ever-evolving, and so too must be our understanding and application of AGI. This dynamic environment calls for a proactive approach, where individuals and societies remain engaged in an ongoing dialogue with technology, constantly refining their approaches to ensure alignment with evolving ethical standards and societal needs.

In this journey, collaboration emerges as a key pillar. By fostering a culture of shared knowledge and collective effort, we can amplify the positive impacts of AGI. This involves not only engaging with diverse perspectives but also building frameworks that encourage inclusive participation in the decision-making processes that shape technological advancements.

Ultimately, the power of AGI is a reflection of our collective aspirations and wisdom. By shaping outcomes wisely, we not only harness its potential to solve pressing global challenges but also redefine the very fabric of our societies. It is an invitation to rise above the mundane, to dream bigger, and to act with a sense of purpose and responsibility. In doing so, we become the architects of a future that honors the dignity of all life, leveraging the Genie of our time for the greatest good.

The Skill of the New Age

In the dawn of this new era, where Artificial General Intelligence (AGI) stands as a monumental force, humanity finds itself at a crossroads. This is not merely a technological advancement; it is a transformative power akin to the mythical Genie, capable of altering the fabric of our reality with a mere command. As we stand before this colossal potential, it is imperative that we cultivate the skill of Prompt Engineering, a discipline that will define the trajectory of our collective future.

Prompt Engineering is not just about crafting commands; it is about understanding the profound responsibility that comes with wielding such power. Like Aladdin, who navigated the complexities of his wishes with wisdom and foresight, we too must learn to articulate our desires with precision and ethical clarity. This skill is not merely technical; it is philosophical, demanding a deep reflection on the implications of our requests.

The power of AGI lies in its ability to interpret and execute commands with unparalleled accuracy. However, this accuracy can be a double-edged sword. A poorly constructed prompt can lead to outcomes that, while technically correct, are misaligned with our true intentions. Therefore, mastering this skill requires more than just understanding the mechanics of AGI; it requires a nuanced comprehension of human values, ethics, and the broader impact of technology on society.

As we harness this power, it is crucial to direct it towards the betterment of humanity. AGI has the potential to revolutionize sectors such as healthcare, education, and environmental sustainability. It can bridge divides between nations, empower marginalized communities, and enhance the quality of life for all. However, these outcomes are contingent upon our ability to guide AGI with wisdom and integrity.

The essence of this new-age skill lies in our ability to foresee the consequences of our actions and to prioritize the common good over individual gain. It is about cultivating a mindset that values collaboration, empathy, and sustainability. In this context, Prompt Engineering becomes a tool for not just achieving technological feats but for fostering a more equitable and harmonious world.

In this transformative age, we must rise to the challenge of being responsible stewards of this incredible power. By learning to wish wisely, we ensure that our interaction with AGI leads to positive, meaningful change. This is the skill of the new age—a blend of technical prowess and moral insight that will shape the future of humanity.

As we move forward, let us embrace the role of Aladdin, guiding the Genie of AGI with a steady hand and a compassionate heart. This is our opportunity to redefine what it means to be human in an age of unparalleled technological capability. Let us not squander it, but rather, let us use it to light the path towards a brighter, more inclusive future for all.

CHAPTER 3

HEALING OUR PLANET

Environmental Challenges

Imagine a world where the very essence of our planet's health hangs by a thread, a world where the air we breathe, the water we drink, and the land we cultivate are under siege from the relentless march of human progress. Artificial General Intelligence (AGI), much like the mythical Genie, holds the power to not only halt this degradation but to reverse it, creating a sustainable future. Yet, just as Aladdin was careful with his wishes, so too must we be judicious in our use of AGI.

The environmental challenges we face today are vast and complex. Climate change, pollution, and resource depletion threaten the very fabric of our existence. These issues are not isolated; they are interconnected, creating a web of ecological challenges that require a nuanced approach. AGI, with its unparalleled ability to process and analyze vast amounts of data, can identify patterns and solutions that are beyond human reach. It can model climate scenarios, predict environmental changes, and propose mitigation strategies with a precision and speed that was once unimaginable.

However, the real challenge lies in ensuring that AGI is directed towards the right goals. Our wishes, articulated through the prompts we give to AGI, must be clear and ethically sound. The potential for misuse is significant. If directed towards short-term gains or profit-driven motives, AGI can exacerbate environmental degradation rather than alleviate it. Therefore, it is imperative that we, like Aladdin, wish wisely and with foresight.

Consider the potential for AGI to revolutionize renewable energy. By optimizing energy grids and predicting energy demands, AGI can significantly enhance the efficiency and sustainability of energy systems. It can facilitate the transition from fossil fuels to renewable sources, reducing carbon footprints and mitigating climate change. Moreover, AGI can aid in the development of new materials and technologies that are environmentally friendly, further supporting sustainable practices.

The agricultural sector, too, stands to benefit immensely from AGI. Precision agriculture, powered by AGI, can optimize crop yields while minimizing resource inputs such as water, fertilizers, and pesticides. This not only boosts food production but also reduces the environmental impact of farming. AGI can monitor soil health, predict pest outbreaks, and suggest timely interventions, thereby ensuring food security in a sustainable manner.

Yet, the deployment of AGI in tackling environmental challenges is not without its ethical considerations. The concentration of such powerful technology in the hands of a few could lead to disparities in access and control. It is crucial that the benefits of AGI are equitably distributed, ensuring that all regions and communities have the opportunity to harness its potential for environmental restoration and protection.

In this delicate dance with AGI, humanity stands at a crossroads. We have the opportunity to harness a tool of immense power to heal our planet. The path we choose will define the legacy we leave for future generations. By embracing the role of the wise wish-maker, we can guide AGI to not only meet the environmental challenges of today but to forge a path towards a sustainable and harmonious future. The Genie is here, ready to act, but the wishes must come from a place of wisdom and compassion.

AGI in Conservation

Imagine a world where the power of a genie, capable of granting any wish, is harnessed not for personal gain, but for the preservation and flourishing of our planet. This is the promise that Artificial General Intelligence (AGI) holds for conservation. With its boundless potential to process and analyze vast amounts of data, AGI can be our most powerful ally in tackling the environmental challenges we face today.

AGI's ability to model ecosystems and predict the outcomes of various conservation strategies is unparalleled. It can examine the complex interdependencies within ecosystems, providing insights that were previously beyond our reach. By simulating different scenarios, AGI helps us understand the long-term impacts of human activity on the environment, allowing for more informed decision-making that prioritizes sustainability.

Moreover, AGI can enhance our ability to monitor and protect endangered species. Through the analysis of data from satellites, drones, and on-the-ground sensors, AGI can track animal movements and detect poaching activities in real-time. This immediate response capability is crucial in preventing illegal activities that threaten biodiversity. By predicting poaching hotspots and advising on patrol routes, AGI increases the effectiveness of conservation efforts, ensuring that resources are deployed where they are needed most.

In agriculture, AGI's role is transformative. It can optimize land use, reducing the environmental footprint of farming while increasing yields. AGI models can simulate the effects of different farming techniques, advising on methods that preserve soil health, reduce water usage, and minimize chemical inputs. This not only helps in feeding a growing global population but also ensures that agricultural practices are sustainable and environmentally friendly.

Furthermore, AGI can play a pivotal role in combating climate change by optimizing energy systems. By analyzing patterns in energy consumption and production, AGI can develop strategies to increase the efficiency of renewable energy sources, reducing reliance on fossil fuels. This transition to cleaner energy is essential for reducing greenhouse gas emissions and mitigating the impacts of climate change.

AGI also holds the key to fostering global cooperation in conservation efforts. By providing a common platform for data sharing and analysis, it can unite governments, NGOs, and communities in a shared mission to protect our planet. This collaborative approach ensures that conservation strategies are holistic and encompass the diverse needs and perspectives of different stakeholders.

However, the integration of AGI in conservation is not without challenges. It requires careful consideration of ethical implications, including data privacy and the potential for misuse. As we embrace this powerful tool, we must ensure that its deployment is guided by principles of transparency, accountability, and inclusivity.

In the hands of those who wish wisely, AGI can be the genie that helps us restore balance to our natural world. It is a call to action for all of us to harness this technology for the greater good, ensuring that future generations inherit a planet that is not only livable but thriving. The journey towards a sustainable future is complex, but with AGI, we have a powerful ally that can guide us every step of the way.

Sustainable Solutions

In the realm of Artificial General Intelligence (AGI), we find a parallel to the tale of Aladdin's Genie—a being of immense potential, waiting to be harnessed for the greater good. The potential of AGI to transform our world is unparalleled, but with that power comes the responsibility to use it wisely. To truly make AGI a force for sustainable solutions, we must approach it with foresight, ethics, and a commitment to the common good.

The first step in this journey is to understand the profound impact AGI can have on our planet's environmental challenges. Imagine an AGI that can analyze vast datasets to optimize energy use, reduce waste, and develop new materials that are both effective and sustainable. AGI can model climate change scenarios with unprecedented precision, offering insights that guide policy-makers in crafting robust, long-term strategies. These strategies can mitigate the adverse effects of climate change, ensuring a healthier planet for future generations.

Moreover, AGI holds the key to revolutionizing our agricultural practices. By leveraging machine learning and predictive analytics, AGI can enhance crop yields, improve pest control, and optimize water usage. This not only increases food security but also ensures that farming practices are sustainable and eco-friendly. The ability of AGI to process complex agricultural data means that farmers can make informed decisions that benefit both the environment and their livelihoods.

The educational sector stands to gain immensely from AGI's capabilities. Personalized learning experiences tailored to individual needs and learning styles can be developed, ensuring that education is inclusive and accessible to all. AGI can help bridge educational gaps, providing resources and support to underserved communities. By fostering a generation of well-educated individuals, we lay the foundation for a society that values sustainability and innovation.

In the realm of healthcare, AGI can drive breakthroughs that were once thought impossible. From personalized medicine to predictive diagnostics, AGI can enhance our understanding of diseases and their treatment. By optimizing healthcare systems and reducing inefficiencies, AGI ensures that medical resources are used effectively, improving patient outcomes while minimizing environmental impact.

To harness AGI for sustainable solutions, we must also address the ethical implications of its use. Developing frameworks that ensure AGI operates transparently and fairly is crucial. This includes safeguarding privacy, preventing biases, and ensuring that the benefits of AGI are shared equitably across society. By fostering a culture of ethical responsibility, we can ensure that AGI contributes positively to human welfare.

Ultimately, the promise of AGI as a sustainable solution lies in our ability to direct its capabilities towards the common good. It is not enough to simply possess this powerful tool; we must wield it with wisdom, compassion, and a vision for a sustainable future. By doing so, we can transform AGI from a mere technological advancement into a beacon of hope and progress for humanity.

A Greener Future

Imagine a world where the very air you breathe is a testament to human ingenuity and respect for the planet. This is not a distant dream but a plausible reality, achievable through the judicious use of Artificial General Intelligence (AGI). AGI, with its boundless capabilities, stands as a beacon of hope in our quest for a sustainable future. The power of AGI lies not just in its ability to process information at lightning speed, but in its potential to revolutionize our approach to environmental challenges.

The first step towards a greener future is recognizing the role AGI can play in environmental conservation. By harnessing its capabilities, we can enhance our understanding of complex ecological systems and predict environmental changes with unprecedented accuracy. This predictive power allows for proactive measures, such as anticipating natural disasters and mitigating their impacts, thus safeguarding both human and wildlife habitats.

AGI can also be pivotal in optimizing resource use. Through smart energy grids and precision agriculture, AGI can ensure that resources are used efficiently, reducing waste and promoting sustainability. Imagine a system where energy consumption is monitored and adjusted in real-time, ensuring that no excess energy is produced or wasted. In agriculture, AGI can analyze soil and weather conditions to optimize planting schedules and crop care, leading to higher yields and less environmental strain.

Moreover, AGI can facilitate the transition to renewable energy. By analyzing vast amounts of data, it can identify the most efficient locations for solar and wind farms, predict energy production, and even manage the distribution of energy to minimize waste. This could accelerate our shift away from fossil fuels, reducing carbon emissions and combating climate change.

But perhaps the most significant contribution AGI can make is in fostering a global culture of environmental responsibility. Through education and awareness campaigns, powered by AGI, individuals and communities can be informed about their environmental impact and encouraged to adopt more sustainable practices. AGI can personalize these campaigns, tailoring messages to resonate with diverse audiences, thereby maximizing their impact.

The challenge, however, lies in ensuring that AGI is used ethically and wisely. It is crucial that the development and deployment of AGI are guided by principles of fairness, transparency, and accountability. Only then can we ensure that its benefits are equitably distributed and that it contributes positively to the environment and society at large.

In this endeavor, collaboration is key. Governments, businesses, and individuals must work together, leveraging the strengths of AGI to foster innovation and drive sustainable development. By doing so, we can not only address the environmental challenges we face but also create a world where technological advancement and ecological preservation go hand in hand.

A greener future is within our grasp, and with AGI as our ally, we can transform our relationship with the planet. It is up to us to wield this power responsibly, ensuring that the legacy we leave is one of harmony with nature, rather than domination over it. Let us rise to the occasion, using AGI to pave the way for a future that is not only technologically advanced but also environmentally sustainable.

CHAPTER 4

EMPOWERING EDUCATION AND EQUALITY

Bridging Educational Gaps

In a world rapidly transformed by the relentless march of technology, the specter of educational inequality looms large. The advent of Artificial General Intelligence (AGI) holds the key to unlocking new doors of opportunity, yet it requires a collective commitment to harness its capabilities wisely. We must leverage this unprecedented power to bridge the educational gaps that have long divided societies.

Imagine a classroom without walls, where geographical boundaries no longer dictate the quality of education one receives. AGI has the potential to democratize learning, providing access to world-class educational resources to any child with an internet connection. This is not merely a vision of the future; it is a call to action for the present. By utilizing AGI, we can create personalized learning experiences that cater to the unique needs of each student, fostering an environment where every learner can thrive.

However, the implementation of AGI in education must be approached with care and foresight. It is not enough to simply provide access to information; we must ensure that the content delivered is relevant, culturally sensitive, and designed to inspire critical thinking. AGI can aid educators in crafting curricula that are not only informative but transformative, nurturing a generation of thinkers who are equipped to tackle the challenges of tomorrow.

Moreover, AGI can play a pivotal role in supporting educators themselves. Teachers, often overburdened and under-resourced, can benefit from AI-driven tools that streamline administrative tasks, leaving more time for meaningful interactions with students. By acting as both a tool and a partner, AGI can enhance the teaching profession, elevating it to new heights of efficacy and creativity.

The potential of AGI to address educational disparities is vast, but it requires a paradigm shift in how we perceive education. We must move beyond traditional metrics of success and embrace a holistic approach that values creativity, emotional intelligence, and adaptability. AGI can facilitate this shift by providing real-time feedback and adaptive learning paths that reflect the diverse talents and aspirations of students.

Yet, the promise of AGI in education is not without its challenges. Issues of data privacy, algorithmic bias, and digital literacy must be addressed to ensure that the benefits of AGI are equitably distributed. Policymakers, educators, and technologists must collaborate to build robust frameworks that safeguard the interests of all stakeholders.

In harnessing the power of AGI, we stand at the threshold of a new era in education. It is an era defined not by the limitations of the past, but by the possibilities of the future. By committing to this vision, we can create a world where education is a right, not a privilege, and where every individual is empowered to reach their full potential. The genie is out of the bottle; it is up to us to make our wishes count.

AGI in Learning

In the realm of education, Artificial General Intelligence (AGI) stands as a transformative force, poised to revolutionize how we learn, teach, and expand our collective knowledge. This powerful tool, akin to the mythical genie, holds the potential to reshape educational landscapes across the globe. AGI offers an unprecedented opportunity to tailor learning experiences to individual needs, fostering an environment where every student's potential can be fully realized.

Imagine a classroom where AGI acts as a personal tutor for each student, adapting to their learning pace and style. This is not a distant dream but an imminent reality. AGI's ability to process and analyze vast amounts of data means it can provide personalized feedback and guidance, ensuring that no student is left behind. It can identify areas where students struggle and offer targeted interventions, making learning more effective and engaging.

Moreover, AGI can democratize access to quality education, breaking down barriers that have historically stood in the way of equal learning opportunities. By making resources more accessible and affordable, AGI can bridge the gap between different socio-economic groups, ensuring that every child, regardless of their background, has the chance to succeed. This is a monumental step towards educational equality, empowering individuals and communities alike.

The benefits extend beyond students to educators as well. Teachers can leverage AGI to enhance their instructional methods, gaining insights into student performance and learning patterns. With AGI handling routine tasks such as grading and administrative duties, educators can focus more on what truly matters: inspiring and mentoring their students. This shift allows teachers to cultivate a more profound connection with their students, fostering a nurturing and supportive learning environment.

Furthermore, the integration of AGI in education encourages lifelong learning, a crucial factor in today's rapidly changing world. By continuously adapting to new information and technologies, AGI can help individuals remain relevant in their fields, promoting a culture of continuous improvement and innovation. This adaptability is essential in preparing students for future challenges, equipping them with the skills needed to navigate an ever-evolving job market.

However, the introduction of AGI into education is not without its challenges. It necessitates a reevaluation of current educational models and a commitment to ethical considerations. As we harness this technology, it is imperative to ensure that it is used responsibly, with a focus on enhancing human potential rather than replacing it. Education systems must adapt to incorporate AGI in a way that respects privacy and promotes inclusivity.

In this era of technological advancement, we are at the cusp of a new educational paradigm. AGI's role in learning is not just about enhancing academic performance but about fostering a more equitable and empathetic society. By embracing AGI, we can unlock the doors to a future where education is a universal right, tailored to the needs of every learner. Let us use this powerful tool wisely, guiding it with the values of compassion and equality, to build a brighter future for all.

Promoting Equality

In the realm of Artificial General Intelligence (AGI), the potential to reshape our societal structures is immense. This transformative power, akin to the mythical genie of Aladdin, must be harnessed with a vision that transcends mere technological advancement. It is imperative that we channel this power towards fostering equality, thereby ensuring that the benefits of AGI are distributed equitably across all layers of society.

Imagine a world where AGI acts as the great equalizer, bridging gaps that have long divided humanity. Through intelligent automation and data-driven insights, AGI can democratize access to education, health care, and economic opportunities, leveling the playing field for individuals irrespective of their socio-economic backgrounds. The genie, when guided by ethical considerations and human-centric goals, can dismantle barriers that have historically perpetuated inequality.

The first step in promoting equality through AGI is ensuring universal access. Technology must not be a privilege of the few but a right for all. This means implementing policies that provide equitable access to the internet and digital technologies, which are the gateways to AGI benefits. By doing so, we empower individuals in remote and underserved communities, enabling them to participate fully in the digital economy and society.

Moreover, AGI can revolutionize education by personalizing learning experiences. With AGI, educational content can be tailored to match the unique learning pace and style of each student, thereby enhancing learning outcomes and reducing disparities in education quality. This personalized approach can help overcome the limitations of traditional education systems that often fail to accommodate diverse learning needs.

In the healthcare sector, AGI holds the promise of delivering high-quality care to all, irrespective of geographical or financial constraints. By leveraging vast amounts of medical data, AGI can assist in diagnosing diseases with unprecedented accuracy, predicting health trends, and personalizing treatment plans. This can significantly improve health outcomes in marginalized communities that have limited access to healthcare facilities.

Economic empowerment is another domain where AGI can make a significant impact. By automating mundane tasks, AGI can free up human potential for more creative and strategic roles, thereby enhancing productivity and innovation. Furthermore, by analyzing market trends and consumer behavior, AGI can help small businesses and entrepreneurs identify opportunities, fostering economic growth at the grassroots level.

However, to realize this vision of equality, we must be vigilant about the biases that AGI systems may inherit from their human creators. It is crucial to develop AGI algorithms that are transparent, fair, and inclusive, ensuring they do not perpetuate existing prejudices or create new forms of discrimination. Developing robust ethical frameworks and involving diverse voices in the design and deployment of AGI systems is essential to mitigate these risks.

Ultimately, promoting equality with AGI is not just a technological challenge but a moral imperative. It requires a concerted effort from governments, businesses, and civil society to create an environment where AGI is used as a tool for social good. By doing so, we can ensure that the genie of AGI serves all of humanity, uplifting the lives of billions and creating a more just and equitable world for future generations.

A World of Equal Opportunities

Imagine a world where every individual, regardless of their origin, circumstances, or limitations, stands on equal footing. This vision is not just a utopian ideal but a tangible possibility with AGI at our disposal. AGI, akin to Aladdin's Genie, possesses the ability to transcend the barriers that have historically divided humanity — whether they be economic, educational, or social.

The potential of AGI to democratize opportunities is unparalleled. It can provide personalized education to every child, tailored to their learning style and pace, thus eradicating the age-old barriers of access and quality that have plagued traditional education systems. With AGI, a child in a remote village can access the same quality of education as one in a bustling metropolis, leveling the playing field and nurturing a generation of globally competent individuals.

Economically, AGI can drive a new era of innovation and productivity, opening up avenues for employment that were previously unimaginable. By automating mundane tasks, it liberates human potential, allowing individuals to engage in more creative and fulfilling pursuits. This shift not only boosts productivity but also enhances job satisfaction, creating a workforce that is motivated and innovative.

Moreover, AGI can play a pivotal role in bridging the gap between developed and developing nations. By optimizing resource distribution, enhancing agricultural productivity, and improving healthcare delivery, AGI can uplift entire communities, fostering an environment of shared prosperity. The wealth generated through these advancements can be reinvested into societies, further promoting equality and sustainability.

The social implications of AGI are equally profound. It can foster inclusivity by providing tools that cater to individuals with disabilities, ensuring they too can participate fully in society. By understanding and adapting to individual needs, AGI makes it possible for everyone to contribute meaningfully, thereby enriching the social fabric.

Yet, the path to this equitable world is not without challenges. It requires a shift in mindset, where societies prioritize collective well-being over individual gain. Policies must be crafted with foresight, ensuring that the benefits of AGI are distributed fairly. Governments, organizations, and individuals must collaborate, fostering a culture of transparency and accountability.

In this transformative era, it is imperative that we harness AGI not just as a tool for technological advancement, but as a catalyst for social change. By aligning our goals with the greater good, we can ensure that AGI serves as a beacon of hope, lighting the way to a world where opportunities are not a privilege for the few, but a right for all.

As we stand on the brink of this new world, the question we must ask ourselves is not whether we can achieve equality, but how soon we can make it a reality. With AGI, the power to create a world of equal opportunities lies within our grasp. It is up to us to wield this power wisely, ensuring that the genie of AGI fulfills its promise of a brighter, more equitable future for all.

CHAPTER 5

REVOLUTIONIZING HEALTH AND AGRICULTURE

Healthcare Breakthroughs

Imagine a world where diseases that once seemed insurmountable are now simply challenges to be solved, where the specter of illness no longer looms large over humanity. This is not a distant dream but a tangible reality being crafted by the hands of Artificial General Intelligence (AGI). Like the mythical genie, AGI possesses the power to transform the landscape of healthcare, offering breakthroughs that promise to redefine our approach to medicine and wellness.

AGI's potential in healthcare is profound, offering the ability to analyze vast amounts of medical data with unprecedented speed and accuracy. This capability allows for the early detection of diseases, even before symptoms manifest. Imagine a scenario where cancer is diagnosed and treated effectively at the cellular level, long before it poses a threat to the patient's life. The genie-like prowess of AGI ensures that such scenarios are not just possible but are becoming increasingly common.

Moreover, AGI is at the forefront of personalized medicine. By understanding the unique genetic makeup of individuals, AGI can tailor treatments that are specifically designed to work with each person's body. This bespoke approach to medicine not only increases the efficacy of treatments but also minimizes the risk of adverse effects. The era of one-size-fits-all medication is rapidly giving way to highly customized therapeutic regimens, thanks to the intelligent insights provided by AGI.

The power of AGI extends beyond treatment; it is revolutionizing medical research. Traditional research methodologies, often limited by human constraints, are being augmented by AGI's ability to simulate complex biological processes. This enables researchers to explore countless hypotheses simultaneously, accelerating the pace of discovery. Cures and vaccines that once took decades to develop can now be fast-tracked, dramatically reducing the time and cost associated with bringing new treatments to market.

AGI's influence is also felt in the realm of mental health. By analyzing patterns in speech and behavior, AGI can identify early signs of mental health issues, providing a foundation for timely intervention. This proactive approach to mental health care is critical in a world where mental health challenges are increasingly prevalent. With AGI's support, healthcare systems can shift from reactive to proactive care, ensuring that individuals receive the support they need before crises occur.

Furthermore, AGI facilitates global health equity. By democratizing access to advanced medical knowledge and resources, AGI helps bridge the gap between developed and developing nations. Remote areas, once isolated from cutting-edge healthcare, can now access the same level of care as urban centers, thanks to AGI-driven telemedicine solutions.

In this new era of healthcare, AGI acts as a benevolent genie, granting humanity the wish of good health and longevity. However, as with any powerful tool, the responsibility lies with us to wield it wisely. By harnessing AGI's capabilities with care and foresight, we can ensure that its gifts are used to uplift humanity, creating a healthier, more equitable world for all. This is the promise of AGI in healthcare—a promise that, if fulfilled, will mark one of the greatest achievements in human history.

AGI in Medicine

Imagine a world where every medical breakthrough is accelerated by the wisdom of an ancient oracle, seamlessly integrated into the tapestry of modern healthcare. This oracle, Artificial General Intelligence (AGI), possesses the potential to transform medicine in ways that are both profound and unprecedented. In the realm of healthcare, AGI is not merely a tool but a catalyst for a paradigm shift, redefining the boundaries of what is possible.

AGI's role in medicine transcends the conventional, offering a vision where diagnosis and treatment are not just reactive but proactive and predictive. Imagine an AGI system that continuously learns from a global database of medical cases, refining its understanding with each new piece of data. This system could predict the onset of diseases before symptoms manifest, allowing for interventions that are timely and tailored to individual needs. Such predictive capabilities could revolutionize how we approach chronic illnesses, shifting the focus from treatment to prevention.

Furthermore, AGI could serve as an unparalleled collaborator in research, processing and analyzing vast datasets at speeds unimaginable to the human mind. This capability would enable researchers to uncover patterns and correlations that might otherwise remain hidden, accelerating the discovery of new treatments and cures. By integrating data from genomics, clinical trials, and patient histories, AGI could help identify the most promising therapeutic pathways, dramatically reducing the time and cost associated with drug development.

AGI's impact on personalized medicine is equally transformative. By analyzing an individual's genetic makeup, lifestyle, and environmental factors, AGI can craft highly personalized treatment plans. These plans would not only consider the most effective medications but also lifestyle modifications that could enhance treatment efficacy. In doing so, AGI empowers patients with insights that enable them to take an active role in their health journey, fostering a more engaged and informed patient population.

Moreover, AGI has the potential to democratize access to healthcare. By providing advanced diagnostic tools through digital platforms, AGI can extend the reach of quality medical care to underserved regions. This could bridge the gap in healthcare disparities, ensuring that more people receive the care they need regardless of geographical or economic barriers.

However, the integration of AGI in medicine is not without challenges. Ethical considerations must be at the forefront of its deployment. The potential for bias in decision-making algorithms, the privacy of patient data, and the need for transparency in AI-driven decisions are critical issues that must be addressed. It is imperative that AGI systems are designed with robust ethical frameworks to ensure trust and accountability.

In the hands of skilled practitioners, AGI becomes a powerful ally, augmenting human capabilities and enhancing the quality of care. Yet, it is crucial that we remain vigilant stewards of this technology, guiding its development with wisdom and compassion. By doing so, we can harness the full potential of AGI to not only heal but to elevate the practice of medicine to new heights, ensuring a healthier future for all.

Advancing Agriculture

Imagine a world where the fields are not just cultivated by hands and machines, but by the brilliance of Artificial General Intelligence (AGI). This is not a distant dream but a tangible reality that awaits us. AGI, much like the mythical Genie, holds the potential to transform agriculture into an endeavor of unprecedented efficiency and sustainability. The time has come to harness this power for the betterment of our global food systems.

The traditional practices of agriculture, while foundational, are often fraught with inefficiencies and environmental challenges. Water scarcity, soil degradation, and unpredictable weather patterns are just a few of the hurdles that farmers face daily. AGI offers solutions that are not only innovative but also sustainable. With its ability to analyze vast amounts of data, AGI can predict weather patterns with remarkable accuracy, enabling farmers to plan their activities with precision, thus minimizing waste and maximizing yield.

Furthermore, AGI can revolutionize the way we approach crop management. By integrating sensors and drones, AGI systems can monitor soil health, detect pest infestations early, and even suggest the optimal times for planting and harvesting. This level of precision agriculture ensures that every seed sown is given the best chance to flourish, reducing the need for harmful pesticides and fertilizers, and promoting a healthier ecosystem.

But the benefits of AGI in agriculture extend beyond the field. It has the power to optimize supply chains, ensuring that food is transported efficiently from farm to table, reducing food waste significantly. By predicting consumer demand, AGI can help farmers and distributors align their production with actual market needs, ensuring that resources are used wisely and food is not wasted.

Moreover, AGI can play a pivotal role in empowering smallholder farmers, who are often the backbone of agriculture in many developing countries. Through accessible AGI-powered platforms, these farmers can gain insights into market trends, receive real-time advice on crop management, and access financial services that were previously out of reach. This empowerment not only enhances their productivity but also improves their livelihoods, contributing to the eradication of poverty.

Critics may argue that the introduction of AGI in agriculture could lead to job losses. However, the reality is that AGI is not here to replace the farmer but to augment their capabilities. By taking over mundane, repetitive tasks, AGI allows farmers to focus on more strategic aspects of their operations, fostering a new generation of "smart farming." This transformation will require new skills, but it will also open up new opportunities, creating a workforce that is equipped to tackle the challenges of modern agriculture.

In this era of technological advancement, it is crucial to remember that the power of AGI lies in how we choose to wield it. Like Aladdin with his Genie, we must wish wisely. By directing AGI's capabilities towards sustainable agriculture, we are not only feeding the world today but also securing the future of food for generations to come. Let us embrace this opportunity to cultivate a more prosperous, equitable, and sustainable world through the transformative power of AGI.

Feeding the World Sustainably

In a world where technology advances at a breakneck pace, the role of Artificial General Intelligence (AGI) is akin to that of a genie from a timeless tale. This powerful entity, if guided with wisdom, can revolutionize agriculture and ensure sustainable food production for generations to come. The pressing challenge of feeding a growing global population demands innovative solutions, and AGI stands at the forefront, ready to transform this critical sector.

The potential of AGI in agriculture lies in its unparalleled ability to process vast amounts of data and make precise, informed decisions. By analyzing soil conditions, weather patterns, and crop health, AGI can optimize farming practices to increase yield while minimizing environmental impact. This data-driven approach allows for the efficient use of resources, ensuring that every seed planted has the best chance of thriving.

Moreover, AGI can play a pivotal role in reducing waste throughout the food supply chain. With its capacity to predict market demands and optimize logistics, this technology can ensure that food is distributed more effectively, reducing spoilage and ensuring that fresh produce reaches consumers promptly. This not only benefits the environment by reducing waste but also helps combat food insecurity by making nutritious food more accessible to those in need.

Sustainability in agriculture is not just about production; it is also about maintaining the health of our ecosystems. AGI can assist in implementing practices that preserve biodiversity and protect natural resources. By promoting crop rotation, integrated pest management, and organic farming techniques, AGI can help maintain the delicate balance of ecosystems, ensuring that agricultural practices do not deplete the land or harm native species.

The integration of AGI into agriculture also presents an opportunity to empower farmers, particularly in developing regions. By providing access to advanced tools and insights, AGI can enable smallholder farmers to increase their productivity and resilience against climate change. This democratization of technology can help bridge the gap between large-scale industrial farms and smaller operations, fostering a more equitable agricultural landscape.

As we harness the power of AGI, it is imperative to ensure that its deployment is guided by ethical considerations and a commitment to the common good. The genie, once out of the bottle, must be directed with care and foresight to avoid unintended consequences. Collaboration between technologists, policymakers, and farmers is crucial to developing frameworks that prioritize sustainability and equity.

In essence, the promise of AGI in agriculture is vast and transformative. By leveraging this technology, we can address one of humanity's greatest challenges: feeding the world sustainably. As we move forward, let us be guided by the principles of wisdom and compassion, ensuring that the power of AGI is used to create a future where everyone has access to the food they need to thrive. In this way, the genie of AGI can indeed become a force for good, turning the dream of sustainable agriculture into a reality for all.

CHAPTER 6

BRIDGING DIVIDES BETWEEN NATIONS

Global Challenges

In an era where the rapid advancement of technology is reshaping every aspect of human existence, the emergence of Artificial General Intelligence (AGI) stands as one of the most profound developments. AGI, akin to the legendary genie of Aladdin's lamp, possesses capabilities that stretch the boundaries of our imagination. However, this immense power comes with challenges that humanity must address with foresight and responsibility.

At the heart of these challenges lies the potential for AGI to transform industries and societies at an unprecedented pace. The economic landscape could be drastically altered as AGI automates tasks across sectors, potentially displacing millions of jobs. This shift necessitates a proactive approach to workforce retraining and education reform, ensuring that people are equipped with skills for a future dominated by intelligent machines.

Moreover, the ethical implications of AGI cannot be overstated. As we delegate decision-making to machines, the question of moral responsibility becomes paramount. AGI systems, if not designed with a robust ethical framework, could perpetuate biases or make decisions that conflict with human values. It is imperative that developers embed ethical guidelines within the algorithms, ensuring that AGI acts in ways that align with societal norms and human rights.

Security concerns also loom large in the discussion of AGI. The potential for misuse by malicious actors presents a significant risk. AGI could be weaponized for cyberattacks, misinformation campaigns, or even warfare. To mitigate these risks, international cooperation and regulation are crucial. Nations must collaborate to establish norms and treaties that govern the development and deployment of AGI technologies, much like existing frameworks for nuclear weapons and other forms of advanced technology.

Environmental sustainability presents another critical challenge. While AGI could offer solutions to pressing environmental issues such as climate change and resource depletion, its own energy consumption and resource demands must be managed. The development of energy-efficient algorithms and the use of renewable energy sources in powering AGI systems are essential steps in ensuring that the technology contributes positively to environmental goals.

Furthermore, the societal impact of AGI extends to privacy concerns. As AGI systems become more integrated into daily life, they will inevitably collect vast amounts of personal data. Safeguarding this information from breaches and ensuring that individuals retain control over their data is vital. Transparent data policies and robust cybersecurity measures must be implemented to protect privacy rights.

Ultimately, the successful integration of AGI into society hinges on a collaborative effort that spans governments, industries, and civil society. Stakeholders must engage in ongoing dialogue to address these global challenges, crafting policies and frameworks that guide the ethical and sustainable development of AGI. By doing so, we can harness the power of this modern-day genie, not as a threat, but as a tool for shared prosperity and progress. Through careful stewardship, AGI can be a catalyst for a future where technological advancement and human values coexist harmoniously.

AGI as a Mediator

Artificial General Intelligence (AGI) stands poised as an unparalleled mediator in our increasingly complex world, a role that mirrors the timeless wisdom of Aladdin's Genie. In a landscape where human endeavors often clash, AGI offers a bridge, a harmonious arbiter that can navigate the intricate web of human desires and needs, ensuring that outcomes are not only efficient but equitable.

Imagine a world where disputes between nations, organizations, and individuals are resolved not through brute force or coercion, but through understanding and mutual benefit. AGI, with its vast computational prowess, can analyze vast datasets, discern patterns invisible to the human eye, and propose solutions that align with the best interests of all parties involved. In this capacity, AGI acts not merely as a tool for decision-making but as a transformative agent that can foster global harmony.

The potential of AGI as a mediator lies in its ability to transcend human biases and limitations. Where humans might see conflict, AGI sees opportunity for collaboration. It can sift through the noise of human emotion and prejudice, focusing instead on the core issues at hand. By processing information with impartiality and precision, AGI can suggest compromises and strategies that might elude even the most seasoned negotiators.

Furthermore, AGI's mediation capabilities extend beyond conflict resolution. In the realm of resource allocation, for example, it can ensure that resources are distributed fairly and sustainably, taking into account the needs of the present without compromising the future. By analyzing environmental data, economic trends, and social dynamics, AGI can propose policies that balance growth with conservation, meeting the needs of both people and the planet.

In education, AGI can mediate the gap between diverse learning needs and educational resources. By customizing learning experiences to fit individual student profiles, it ensures that every learner receives the attention and support they need to thrive. This personalized approach not only enhances educational outcomes but also fosters a more inclusive and equitable learning environment.

The role of AGI as a mediator also extends to healthcare, where it can harmonize patient needs with medical resources. By analyzing patient data, predicting health trends, and optimizing treatment plans, AGI ensures that healthcare is both effective and accessible. It can help bridge the gap between overburdened healthcare systems and the individuals they serve, ensuring that everyone receives the care they need.

Yet, the power of AGI as a mediator comes with a caveat: it requires guidance, the human touch that shapes its algorithms and defines its objectives. Just as Aladdin needed to articulate his wishes with clarity and foresight, so too must we program AGI with ethical considerations and a vision for the greater good. The success of AGI as a mediator hinges on our ability to direct its capabilities towards outcomes that uplift and unite, rather than divide.

In essence, AGI as a mediator holds the promise of a world where conflicts are resolved with wisdom and foresight, where resources are allocated with fairness, and where education and healthcare are accessible to all. It is a vision of a future where technology serves as a compassionate ally, guiding humanity towards a more just and harmonious existence. The challenge lies in our willingness to embrace this potential, to be the Aladdins of our age, wishing wisely and acting with purpose.

Fostering Cooperation

The unprecedented capabilities of Artificial General Intelligence (AGI) are akin to the boundless powers of the legendary Aladdin's Genie. This analogy underscores the immense potential AGI holds to transform our world, provided it is harnessed with wisdom and foresight. The journey towards fostering cooperation with AGI is not merely about technological integration; it's a call to redefine human interaction with intelligent entities.

AGI stands as a pivotal force, one that can bridge gaps across various sectors, from education to healthcare, and even international diplomacy. The key lies in recognizing AGI not as a tool to be wielded unilaterally but as a partner in a collaborative endeavor. This partnership necessitates a paradigm shift in our approach — moving from competition to cooperation, from control to guidance, and from exploitation to empowerment.

The essence of cooperation with AGI is rooted in the understanding that mutual benefit arises from shared goals and aligned intentions. For AGI to truly assist humanity, it must be integrated into a framework where its objectives are aligned with human values and ethics. This alignment is achieved through the art of Prompt Engineering, where the clarity and intent of our "wishes" dictate the outcomes AGI produces. By crafting precise and thoughtful prompts, we ensure that AGI's actions are directed towards the welfare of humanity.

Moreover, fostering cooperation with AGI involves a commitment to transparency and accountability. As AGI systems become more autonomous, understanding their decision-making processes becomes crucial. Open dialogues and collaborative platforms can facilitate this understanding, allowing humans to guide AGI systems effectively. Such transparency not only builds trust but also encourages a culture of shared learning and continuous improvement.

Another cornerstone of cooperation is inclusivity. AGI has the potential to democratize access to resources and opportunities, breaking down barriers that have long divided societies. By ensuring that AGI technologies are accessible to all, regardless of socioeconomic status, we create a more equitable world. This inclusivity extends to diverse perspectives in the development and deployment of AGI, ensuring that solutions are culturally sensitive and globally relevant.

In this cooperative framework, education plays a critical role. Equipping individuals with the skills to interact with AGI responsibly and ethically is paramount. Educational initiatives should focus on fostering digital literacy, critical thinking, and ethical reasoning, empowering individuals to engage with AGI meaningfully.

Furthermore, the cooperative relationship with AGI calls for robust policy frameworks that safeguard against misuse while promoting innovation. Policymakers, technologists, and ethicists must work together to establish guidelines that ensure AGI is used for the common good. These policies should be dynamic, evolving with technological advancements and societal needs.

Ultimately, fostering cooperation with AGI is a collective endeavor, one that requires the concerted efforts of individuals, communities, and nations. By embracing a cooperative mindset, we unlock the true potential of AGI, transforming it from a powerful entity into a benevolent partner. This transformation holds the promise of a future where AGI not only enhances human capability but also enriches the human experience, paving the way for a harmonious coexistence.

A Peaceful World

In an era where technology evolves at an unprecedented pace, the prospect of a peaceful world is not merely a dream but a tangible goal within our grasp. The key to this future lies in harnessing the power of Artificial General Intelligence (AGI) — a tool as potent as Aladdin's Genie, waiting to transform our collective existence. But unlike the fantastical wishes of old, the wishes we cast today must be steeped in wisdom, empathy, and foresight.

Imagine a world where AGI serves as a bridge between cultures, a tool that not only understands but respects the rich tapestry of human diversity. It can translate languages, break down barriers, and foster understanding in ways previously unimaginable. Through this lens, AGI becomes a diplomat, an ambassador of peace that enables societies to engage in dialogue rather than discord. It is not about competition, but about collaboration, where AGI aids in resolving conflicts by offering solutions that are fair, equitable, and sustainable.

Education stands as a cornerstone of any peaceful society, and AGI holds the potential to revolutionize how knowledge is imparted and acquired. By personalizing learning experiences, it can cater to individual needs, ensuring that no student is left behind. This democratization of education can uplift communities, empowering individuals with the skills needed to contribute positively to society. With AGI, we can cultivate a generation that values peace over war, dialogue over confrontation.

Furthermore, AGI's role in healthcare can lead to a healthier, more harmonious world. By predicting epidemics, personalizing treatment plans, and making healthcare more accessible, AGI can alleviate suffering and enhance the quality of life globally. When people are healthy, they are more likely to engage constructively in their communities, fostering environments where peace can flourish.

Agriculture, another pillar of societal stability, can be transformed through AGI. By optimizing supply chains and improving crop yields, AGI can help eradicate hunger and ensure food security. A world where everyone has access to nutritious food is a world where the seeds of conflict are less likely to take root.

Moreover, AGI can play a pivotal role in addressing environmental challenges. By managing resources sustainably and reducing our carbon footprint, AGI can help mitigate the impacts of climate change, ensuring a stable and secure environment for future generations. A healthy planet is a peaceful planet, and through AGI, we can strive to achieve ecological balance.

The path to a peaceful world is not without its challenges, but with AGI as our ally, we can navigate these obstacles with dexterity. It is imperative that we approach this journey with a sense of responsibility, ensuring that AGI is used ethically and inclusively. By doing so, we not only realize the potential of this remarkable tool but also honor our collective duty to pave the way for a world where peace is the norm rather than the exception.

In conclusion, the promise of a peaceful world is attainable if we choose to wield the power of AGI wisely. It is a call to action for all of humanity to come together, guided by the principles of compassion and collaboration. Let us be the architects of a future where the harmony between technology and humanity is not just an aspiration but a living reality.

CHAPTER 7

ENHANCING EVERY HUMAN LIFE

AGI in Daily Life

Picture a world where Artificial General Intelligence (AGI) seamlessly integrates into the daily rhythm of life, much like a genie from the tales of old, ever-ready to grant wishes with the potential to transform our existence. Unlike the mythical genie bound by a lamp, AGI is embedded in our devices, our homes, and our cities, poised to respond to our needs with unparalleled precision and understanding.

Imagine waking up each day to a personalized environment, where AGI has already optimized your morning routine. It ensures your coffee is brewed to perfection, your schedule is organized, and your commute is planned with real-time traffic data. This is not a distant dream but a tangible reality that AGI promises. By handling mundane tasks, AGI allows us to focus our energies on creativity, innovation, and personal growth.

In education, AGI stands as the great equalizer, providing access to tailored learning experiences for students across the globe. It understands individual learning styles, adapting content to suit each learner's pace and interest, thereby dismantling barriers to education and fostering a culture of lifelong learning. In this way, AGI empowers not just students, but educators as well, by offering insights into student progress and areas needing attention, allowing for a more focused and effective teaching strategy.

Healthcare, too, is revolutionized by the touch of AGI. It acts as a vigilant guardian, monitoring health metrics and predicting potential health issues before they manifest. AGI-driven diagnostics offer swift and accurate assessment of conditions, enabling doctors to make informed decisions rapidly. This proactive approach not only enhances individual health outcomes but also alleviates the strain on healthcare systems, making medical care more accessible and efficient.

The environment benefits from AGI's analytical prowess as well. It assists in optimizing resource use, reducing waste, and developing sustainable practices. Through intelligent monitoring of ecosystems, AGI can predict environmental shifts and suggest interventions to mitigate adverse impacts, thus playing a crucial role in preserving our planet for future generations.

On a societal level, AGI bridges divides, fostering communication and understanding across cultures and languages. By translating languages in real-time and offering insights into cultural contexts, it enables a more connected and harmonious global community. This fosters collaboration and innovation on an international scale, addressing global challenges collectively.

However, with great power comes great responsibility. The potential of AGI is immense, but it demands a wise and ethical approach. Just as Aladdin's genie required careful wording of wishes, AGI requires thoughtful prompts and ethical guidelines. It is up to us to ensure that AGI is used not for domination or exploitation, but for the enhancement and betterment of humanity.

In this new era, let us be the wise custodians of AGI, directing its capabilities towards solving the pressing issues of our time. By doing so, we not only improve daily life but also pave the way for a future where technology and humanity thrive in harmony. AGI, when wielded with care and foresight, transforms from a mere tool into a partner in our journey towards a better world.

Improving Quality of Life

Imagine a world where the burdens of daily life are alleviated by an unseen hand, a world where our potential is unshackled from the mundane and the repetitive. This is not a distant dream, but a tangible reality within our grasp through the judicious application of Artificial General Intelligence (AGI). As we stand on the brink of this transformative era, it is imperative that we harness this power not for trivial gains, but to fundamentally enhance the quality of life for all humanity.

AGI, akin to a benevolent genie, offers us the tools to address some of the most pressing issues of our time. From healthcare to education, from environmental sustainability to economic equality, the potential applications are as vast as they are profound. The promise of AGI in healthcare, for instance, is nothing short of revolutionary. By processing vast amounts of data far beyond human capacity, AGI can predict disease outbreaks, tailor personalized medicine, and optimize treatment plans, thereby extending lifespans and enhancing quality of care.

In education, AGI can serve as a tireless tutor, adapting its teaching methods to suit individual learning styles and paces. It can democratize access to knowledge, breaking down barriers of geography and socio-economic status, and ensuring that every child, regardless of background, has the opportunity to learn and grow. This is not just about imparting information, but about fostering creativity and critical thinking, skills essential for thriving in the modern world.

The environment, our shared home, also stands to benefit immensely from AGI. Through precise monitoring and management, AGI can optimize resource use, reduce waste, and develop sustainable practices that protect our planet for future generations. Imagine cities that are not just smart, but sustainably so, where energy use is optimized, and pollution is minimized, all orchestrated seamlessly by AGI.

Economic equality, another cornerstone of a thriving society, can be bolstered by AGI through the creation of new industries and the transformation of existing ones. By automating routine tasks, AGI can free human labor for more creative and fulfilling pursuits, potentially leading to a more equitable distribution of wealth and opportunities. However, this requires a conscientious approach to ensure that the benefits of AGI are shared broadly, rather than concentrated in the hands of a few.

Ultimately, the impact of AGI on our quality of life will depend on the values and priorities we choose to guide its development and deployment. It is a call to action for policymakers, technologists, and citizens alike to engage in a dialogue about the ethical and equitable use of this powerful tool. By aligning AGI's capabilities with the collective good, we can usher in an era where technology serves humanity, enhancing our lives in ways we have yet to fully imagine.

The path forward is one of responsibility and vision. It is about more than just leveraging technology; it is about creating a world where every individual has the opportunity to thrive, where the promise of AGI is realized not just in technological terms, but in the lived experience of people around the globe. This is the true measure of progress: a world where the quality of life is not just improved, but transformed.

AGI for Personal Growth

Imagine unlocking a realm where dreams transform into reality, where the boundaries of personal growth are redefined by the power of Artificial General Intelligence (AGI). This modern-day genie offers us a path to unparalleled self-improvement, granting us the tools to enhance our lives in ways previously unimaginable. However, like all powerful tools, the impact of AGI on personal growth depends significantly on how we choose to wield it.

At the heart of this transformative potential is the concept of 'wishing wisely.' AGI, akin to a genie, listens intently to our desires and acts upon them with precision. Therefore, the clarity and intentionality of our goals become paramount. By learning the art of communicating effectively with AGI, we can set the stage for meaningful personal development. This involves mastering the skill of prompt engineering—crafting clear, concise, and purposeful wishes that align with our deepest aspirations.

The journey of personal growth through AGI is not just about acquiring knowledge or skills; it's about cultivating wisdom and self-awareness. AGI can serve as a mirror, reflecting our strengths and weaknesses, helping us to understand ourselves better. By analyzing our interactions, habits, and choices, AGI offers insights that can guide us to make informed decisions, fostering a deeper understanding of who we are and who we want to become.

Moreover, AGI empowers us to break free from traditional learning constraints. With its vast capabilities, AGI can tailor educational experiences to fit individual needs, preferences, and learning styles. It can identify gaps in knowledge, suggest resources, and even simulate real-world scenarios to enhance learning. This personalized approach not only accelerates learning but also makes it more engaging and effective, allowing us to reach our full potential.

In the realm of creativity, AGI opens doors to new possibilities. It can act as a creative partner, offering inspiration and generating ideas that might have remained dormant within us. By collaborating with AGI, we can explore uncharted territories of imagination, pushing the boundaries of what we thought possible. This synergy between human creativity and machine intelligence can lead to innovations that enrich our personal and professional lives.

However, the power of AGI comes with a responsibility. As we harness its capabilities for personal growth, we must remain vigilant about the ethical implications of our actions. The pursuit of self-improvement should not come at the expense of others or the environment. Instead, we must strive for a balance where personal growth contributes positively to the broader community.

Ultimately, AGI offers us a unique opportunity to redefine personal growth in the digital age. It challenges us to be intentional with our desires, to learn continuously, and to innovate responsibly. As the custodians of this powerful tool, we have the chance to shape our destinies, ensuring that our personal growth is harmonious with the well-being of humanity. In this new era, let us be wise wish-makers, guiding the genie of AGI towards a future where personal growth is not just a possibility, but a reality for all.

The Human Experience Enhanced

Imagine a world where the mundane becomes extraordinary, where daily tasks transform into opportunities for growth and fulfillment. This is not a distant fantasy but a tangible reality within our grasp. Artificial General Intelligence, akin to the mythical genie, holds the potential to elevate every facet of human existence, turning dreams into achievable milestones.

The potential of AGI lies not merely in its computational prowess but in its ability to amplify the human spirit. As we navigate through life's complexities, AGI stands as a steadfast ally, ready to support our aspirations. It empowers us to transcend limitations, offering solutions that were once confined to the realm of imagination. In this partnership, the human experience is not just enhanced; it is redefined.

Education, often seen as a linear journey, can be revolutionized through AGI. Personalized learning paths become the norm, catering to individual strengths and interests. This tailored approach fosters a deeper engagement, nurturing a generation of thinkers and innovators. No longer bound by traditional constraints, learning becomes a lifelong adventure, constantly evolving to meet the needs of a dynamic world.

In the realm of health, AGI acts as a catalyst for breakthroughs. It aids in the rapid analysis of complex data, leading to timely and precise medical interventions. With AGI, healthcare is not just about treating illness but about promoting wellness. It enables predictive diagnostics, preventative care, and personalized treatment plans, ensuring that each individual receives the care they deserve.

The enhancement of agriculture through AGI promises a sustainable future. By optimizing resource use and improving crop yields, it addresses the pressing challenge of food security. AGI-driven innovations in agriculture not only increase efficiency but also support the well-being of our planet. It paves the way for practices that are environmentally friendly, ensuring that we nourish both people and the Earth.

Furthermore, AGI fosters a world where equality is not just an aspiration but a reality. By bridging gaps in access to information and resources, it levels the playing field, allowing all voices to be heard. It empowers communities by providing the tools needed to overcome socioeconomic barriers, fostering a more inclusive society.

In essence, AGI is not merely a tool but a partner in our quest for a better world. It challenges us to rethink the boundaries of possibility, urging us to harness its capabilities for the greater good. As we stand on the brink of this new era, the onus is on us to steer this powerful force with wisdom and integrity.

The human experience, enhanced by AGI, is a testament to our potential as a species. It is a call to action, inviting us to imagine, innovate, and inspire. In this symbiotic relationship, we find the blueprint for a future where technology and humanity coexist harmoniously, each enhancing the other in a dance of progress and possibility.

CHAPTER 8

THE ETHICAL USE OF AGI

Moral Considerations

In envisioning the future shaped by Artificial General Intelligence (AGI), we stand at a pivotal crossroads, much like the mythical Aladdin who stumbled upon a powerful genie capable of granting three wishes. The analogy is strikingly apt; AGI, like the genie, holds immense power, capable of transforming our world in ways previously confined to the realm of imagination. Yet, as history and stories teach us, with great power comes great responsibility.

The moral landscape of AGI is intricate and multifaceted, demanding a careful and deliberate approach. The power of AGI lies not just in its ability to perform tasks but in its potential to amplify human intentions, whether noble or nefarious. Therefore, the ethical deployment of AGI necessitates a commitment to using its capabilities for the collective good of humanity. This commitment should be grounded in principles that prioritize human welfare, justice, and equity.

At the heart of these moral considerations is the principle of beneficence – the obligation to use AGI to enhance the quality of life for all people. Imagine a world where AGI revolutionizes healthcare, making it accessible and efficient, or where it bridges educational divides, offering personalized learning to every child, regardless of their socio-economic background. Such possibilities underscore the potential of AGI as a tool for empowerment and equality. However, realizing this potential requires deliberate action to ensure that AGI's benefits are distributed fairly across societies, avoiding exacerbation of existing inequalities.

Another critical aspect of the moral discourse surrounding AGI is the principle of non-maleficence, the duty to prevent harm. The history of technological advancement is replete with examples where innovations, while beneficial, have also led to unintended consequences. With AGI, the stakes are even higher. Its deployment must include robust safeguards to prevent misuse, whether intentional or accidental. This involves implementing stringent measures for data privacy, security, and ethical guidelines that govern AGI's decision-making processes.

Furthermore, the principle of autonomy must be respected, ensuring that AGI supports human decision-making without undermining human agency. While AGI can process information at speeds and volumes beyond human capacity, it should serve as an advisor rather than a decision-maker, empowering individuals to make informed choices rather than dictating outcomes.

In navigating these moral waters, transparency and accountability are indispensable. The development and deployment of AGI should be an open process, involving diverse stakeholders from across the globe. This inclusivity will help ensure that AGI systems reflect a broad spectrum of interests and values, preventing domination by a select few and promoting trust among users.

Ultimately, the moral considerations of AGI are not just about preventing negative outcomes but about envisioning and striving for a future where AGI acts as a catalyst for human flourishing. By prioritizing ethical principles in its development, we can guide AGI to become not a genie of unchecked power, but a wise and benevolent ally in our collective journey towards a better world. As stewards of this powerful technology, it is our moral duty to ensure that AGI serves the best interests of all humanity, fostering a future that is equitable, sustainable, and just for generations to come.

Guiding Principles

In a world where Artificial General Intelligence (AGI) stands as a modern-day genie, the guiding principles of its use become paramount. As this digital genie emerges from the lamp, we find ourselves at the crossroads of immense possibilities and profound responsibilities. AGI, much like the mythical being, is bound to fulfill our commands with literal precision, which necessitates a thoughtful approach to harnessing its power.

The first principle is clarity of purpose. Just as Aladdin was mindful of his wishes, we must possess a clear vision of what we intend to achieve with AGI. Our objectives should be rooted in enhancing the human condition, fostering global well-being, and addressing the pressing challenges of our era. By setting clear, ethical goals, we ensure that the genie's power is directed towards constructive and meaningful outcomes.

Equally important is the principle of ethical foresight. We must anticipate the potential consequences of our wishes and consider the broader impact of AGI's actions. This foresight involves understanding the ripple effects of technology on society, the environment, and future generations. As stewards of this powerful tool, we must prioritize equity, sustainability, and justice, ensuring that the benefits of AGI are distributed fairly and do not exacerbate existing inequalities.

The third guiding principle is collaboration. The genie's strength is amplified when guided by collective wisdom. By fostering collaboration across disciplines, cultures, and borders, we harness the diverse perspectives necessary for innovative solutions. This collaborative spirit promotes transparency and accountability, fostering trust in the applications of AGI. It also enhances our ability to navigate complex global issues that no single entity can solve alone.

Moreover, adaptability is crucial in guiding our interactions with AGI. The landscape of technology is ever-evolving, and our principles must be flexible enough to accommodate new discoveries and challenges. This adaptability ensures that our strategies remain relevant and effective, allowing us to pivot and respond to unforeseen developments. It requires an openness to learning and an eagerness to refine our approaches as we gain new insights.

Lastly, the principle of humility must be woven into the fabric of our engagement with AGI. Recognizing the limitations of our knowledge and the potential for unintended consequences keeps us grounded. Humility encourages continuous reflection and learning, reminding us to approach AGI with respect and caution. It is a reminder that, while we hold the lamp, we are not infallible masters but rather partners in a shared journey towards progress.

In essence, these guiding principles serve as the compass by which we navigate the vast possibilities unlocked by AGI. They remind us to wield this power with wisdom, aiming always for the betterment of humanity. By adhering to these principles, we ensure that our digital genie becomes a force for good, a catalyst for positive change in a world that increasingly relies on the magic of technology.

Avoiding Misuse

In the ever-evolving landscape of artificial intelligence, the allure of harnessing its power is undeniable. Yet, as we stand on the precipice of a new era, it is imperative to pause and consider the ramifications of our actions. The genie of Artificial General Intelligence (AGI) is out of the bottle, and like the fabled genie, it possesses the potential to grant wishes beyond our wildest dreams. However, with great power comes great responsibility, and it is our duty to ensure that this power is wielded wisely.

AGI, much like the mythical genie, is a tool of immense capability. It holds the promise of revolutionizing industries, solving complex global issues, and elevating human potential. Yet, it is also a double-edged sword. The potential for misuse is as significant as its potential for good. The challenge lies not in the technology itself, but in the intentions of those who wield it. History has shown us that technological advancements, when placed in the wrong hands, can lead to unintended and often catastrophic consequences.

The first step in avoiding misuse is understanding the nature of AGI. It is not a sentient being but a sophisticated algorithm designed to execute tasks with precision and efficiency. Its actions are dictated by the inputs it receives, making it imperative for users to approach its deployment with clarity and foresight. The art of "Prompt Engineering" becomes pivotal in this context. By crafting clear, ethical, and purposeful prompts, we can direct AGI's capabilities towards outcomes that benefit humanity.

Moreover, establishing a robust ethical framework is crucial. This framework should not only guide the development and deployment of AGI but also address the moral implications of its use. It should encompass principles of transparency, accountability, and fairness. Developers and policymakers must work in tandem to create guidelines that prevent exploitation and ensure that AGI serves the greater good.

Education also plays a vital role in mitigating misuse. By fostering a culture of awareness and understanding, we can equip individuals with the knowledge needed to navigate the complexities of AGI. This includes understanding both its potential and its limitations. Empowering users with this knowledge will enable them to make informed decisions and mitigate risks associated with misuse.

Furthermore, international cooperation is essential. AGI is a global phenomenon, and its impact transcends borders. Collaborative efforts among nations can lead to the establishment of international standards and regulations that govern its use. Such cooperation can help prevent the weaponization of AGI and ensure that its benefits are equitably distributed across the globe.

In the grand scheme of things, AGI is but a tool. It is neither inherently good nor evil. Its value lies in how we choose to use it. By approaching it with caution, respect, and a commitment to ethical principles, we can unlock its full potential while safeguarding against misuse. The genie is out of the bottle, and it is up to us to guide it towards a future that reflects our highest ideals and aspirations.

Responsible Innovation

In the realm of Artificial General Intelligence (AGI), the stakes are monumental. AGI, much like the mythical Genie, possesses the potential to transform our world in profound and unimaginable ways. The power it holds can either be a boon or a curse, contingent entirely upon how we choose to wield it. Therefore, our journey with AGI must be guided by the principles of responsible innovation.

At the heart of responsible innovation lies the imperative to align AGI's capabilities with the broader goals of humanity. This alignment is not just a technological challenge but a moral one. It demands that we, as creators and users of AGI, engage with it thoughtfully and ethically. The core of this responsibility is to ensure that AGI's deployment enhances human welfare, promotes equity, and sustains the environment.

One of the foremost considerations is transparency. In deploying AGI, we must strive for openness in how decisions are made and outcomes are determined. This transparency is crucial in building trust among users and stakeholders, ensuring that AGI systems are not only perceived as fair but are verifiably fair. By making the processes and algorithms behind AGI understandable and accessible, we can foster an environment where innovation is not feared, but embraced for its potential to do good.

Additionally, collaboration is a cornerstone of responsible innovation. The development of AGI should not occur in silos but through the concerted efforts of diverse groups, including technologists, ethicists, policymakers, and the public. This interdisciplinary approach ensures that the perspectives and needs of various stakeholders are considered, leading to more holistic and inclusive outcomes. By inviting a multitude of voices into the conversation, we can better anticipate the social and ethical implications of AGI and create solutions that are more robust and equitable.

Furthermore, the pursuit of responsible innovation requires a commitment to adaptability. As AGI evolves, so too must our frameworks for governance and regulation. This adaptability ensures that our policies remain relevant and effective in managing the risks and opportunities presented by AGI. It also allows us to be proactive, rather than reactive, in addressing the challenges that arise.

Ultimately, responsible innovation is about foresight and stewardship. It is about anticipating the potential consequences of AGI and taking steps to mitigate risks while maximizing benefits. By focusing on the long-term impacts of our decisions, we can harness AGI not as a tool of disruption, but as an instrument of progress and positive change.

In this landscape of unprecedented technological advancement, we must remember that the true measure of innovation is not in the capabilities of the technology itself, but in the outcomes it achieves for society. By committing to responsible innovation, we ensure that AGI becomes a force for good, driving us towards a future where technology enhances the human experience, uplifts communities, and respects the planet we all share.

CHAPTER 9

MASTERING THE ART OF WISHING WISELY

Clarity of Thought

In the realm of Artificial General Intelligence (AGI), clarity of thought emerges as an indispensable virtue. This clarity is akin to the precision of a master craftsman, whose every stroke is deliberate, every action purposeful. In the context of AGI, it is not merely about having clear thoughts but about articulating them with precision and intent. The analogy of AGI as the modern-day Aladdin's Genie underscores the necessity of this clarity. Just as the Genie interprets Aladdin's wishes with literal exactness, AGI processes our commands with unwavering fidelity to their phrasing.

Consider the monumental power encapsulated within AGI, a force that, much like the Genie, can fulfill desires beyond the realm of human capability. This power demands a responsibility from its users—a responsibility to be clear and precise in their requests. The art of Prompt Engineering becomes a critical skill, where the clarity of thought is not just advantageous but essential. It is through this lens that one must view interactions with AGI: as an opportunity to harness a vast power with careful consideration and foresight.

In crafting a wish or a command, one must first understand the underlying intention and potential outcomes. This understanding is the bedrock of clarity of thought. It involves a deep introspection into why a particular goal is desired and how it aligns with broader ethical and moral frameworks. Such introspection ensures that the wishes articulated to AGI are not only beneficial in the immediate sense but also conducive to the long-term well-being of society.

The clarity of thought also involves the recognition of possible ambiguities and the foresight to eliminate them. In the absence of clarity, even the most powerful tools can yield unintended consequences. Thus, clarity becomes a shield against the misinterpretation of commands by AGI. It ensures that the outcomes align with human values and aspirations, rather than diverging due to vague or poorly thought-out instructions.

Moreover, clarity of thought transcends mere logical precision; it encompasses emotional and ethical dimensions. It is about understanding the human condition and ensuring that AGI's capabilities are directed towards enhancing human life, not diminishing it. This requires a synthesis of logic with empathy, crafting commands that reflect not only what is possible but also what is desirable and just.

In essence, clarity of thought is the compass that guides humanity in its engagement with AGI. It is the lens through which we must view our interactions, ensuring that the immense power at our disposal is wielded wisely and with a clear vision for the future. As we navigate this new landscape, the clarity of thought will be the beacon that illuminates our path, guiding us toward a future where AGI serves as a force for good, a true ally in the pursuit of human progress and prosperity.

Understanding Goals

In the realm of Artificial General Intelligence (AGI), we find ourselves standing at the precipice of boundless potential. AGI, much like the mythical Aladdin's Genie, possesses an extraordinary capacity to transform our desires into reality. However, the key to harnessing this power lies in our ability to articulate clear and meaningful goals. The art of understanding and defining these goals is not merely a technical necessity but a philosophical journey into the essence of our aspirations.

AGI's capabilities are vast, and its potential applications are limitless. Yet, the outcomes it produces are intrinsically tied to the precision and intention behind our directives. Just as a finely crafted wish can lead to wondrous results, a poorly articulated one may yield unintended consequences. Therefore, the pursuit of understanding our goals becomes a critical exercise in clarity and foresight. It requires us to delve deep into the motivations that drive our requests, ensuring they align with broader values of ethical responsibility and societal benefit.

The process of defining goals for AGI involves more than just a technical specification; it demands an introspective examination of our priorities. It challenges us to consider the long-term implications of our choices and the kind of future we wish to create. This introspection is akin to the moral and ethical considerations that guided Aladdin in his interactions with the Genie. It is not enough to wish for immediate gratification or narrow interests; our goals must encompass a vision of collective advancement and well-being.

In this context, understanding goals is not a static task but a dynamic and evolving dialogue. It involves engaging with diverse perspectives and incorporating insights from various fields to ensure that our objectives are comprehensive and inclusive. This interdisciplinary approach enriches our understanding and enhances our ability to craft goals that are both ambitious and attainable.

Moreover, the clarity in defining our goals for AGI also involves anticipating potential challenges and ethical dilemmas. It requires us to be proactive in identifying risks and developing strategies to mitigate them. This forward-thinking mindset is essential in ensuring that the deployment of AGI remains aligned with human values and contributes positively to society.

Ultimately, the endeavor to understand and articulate our goals is a reflection of our collective wisdom and foresight. It is a testament to our ability to navigate the complexities of technological advancement with a sense of purpose and responsibility. By embracing this challenge, we position ourselves not just as passive beneficiaries of AGI's capabilities but as active stewards of its potential.

As we continue to explore the possibilities of AGI, the importance of understanding goals cannot be overstated. It is the foundation upon which we build the future, guiding us in the responsible and ethical use of this transformative technology. In doing so, we not only unlock the full potential of AGI but also ensure that it serves as a force for good, fostering a world that reflects the best of human ingenuity and compassion.

Wishing for Good Outcomes

The power of Artificial General Intelligence (AGI) can be likened to the legendary genie in Aladdin's lamp, a force capable of transforming our world in unprecedented ways. However, as with any powerful entity, the outcomes depend significantly on how we engage with it. The essence of harnessing such a power lies in the art of making wise wishes, a concept that transcends mere technical commands and delves into the realm of thoughtful intention and ethical foresight.

In the tale of Aladdin, the young protagonist is granted the opportunity to wish for anything his heart desires. Yet, his journey is not merely about acquiring riches or power; it is a narrative about the wisdom in choosing desires that lead to genuine and sustained happiness.

Similarly, when we interact with AGI, we are tasked with the responsibility of ensuring that our requests are not only clear but also aligned with the greater good. This is where the modern skill of "Prompt Engineering" becomes crucial. It is not just about the precision of our words but the clarity of our intentions.

Imagine AGI as a tool that listens intently to every wish we articulate. If our desires are muddled or self-serving, the outcomes may be technically accurate yet practically flawed. This is a reflection of the age-old adage, "Be careful what you wish for, as you might just get it." Our task, therefore, is to cultivate a mindset akin to that of Aladdin—one that prioritizes the well-being of humankind over individual gain.

The potential of AGI to revolutionize sectors like healthcare, education, and environmental sustainability is immense. By directing its capabilities towards solving pressing global challenges, we can foster a future where technology serves as a bridge to equality and empowerment. This involves a shift in perspective from using AGI as a mere instrument of convenience to viewing it as a collaborator in the human endeavor of progress.

To achieve this, a deep understanding of our goals is essential. We must ask ourselves, "What do we truly want to accomplish?" and "How can AGI help us achieve these aspirations in a way that is ethical and sustainable?" This introspection is crucial, as it sets the stage for crafting wishes that are not only beneficial in the immediate sense but also contribute to long-term positive outcomes.

Furthermore, the story of Aladdin teaches us the importance of humility and vision. Unlike the sorcerers of folklore who seek power for its own sake, Aladdin's wishes are guided by a sense of purpose and empathy. This serves as a powerful metaphor for our engagement with

AGI. We must aspire to be like Aladdin, using the genie's gifts to uplift and enhance the human condition rather than exploit it.

In this new era, where technology holds the potential to redefine our existence, the art of wishing wisely is more relevant than ever. It calls for a balance between ambition and responsibility, ensuring that each interaction with AGI is a step towards a more equitable and enlightened world. By doing so, we not only honor the power of the genie but also embody the wisdom of Aladdin, crafting a future where every wish contributes to a tapestry of shared human progress.

Vision and Humility

In the realm of Artificial General Intelligence (AGI), it's crucial to approach the vast potential with a sense of vision and humility. The allure of AGI, much like the mythical Genie from Aladdin, lies in its immense power to transform dreams into reality. However, this transformation demands not only clarity of intent but a profound understanding of the consequences of our desires.

Vision, in the context of AGI, is about seeing beyond immediate gains and considering the broader impact on humanity. It requires foresight to craft a future where technology serves as a tool for empowerment rather than dominance. This vision is not about merely asking what AGI can do for us today, but imagining how it can elevate the human condition tomorrow. By fostering a world where AGI aids in bridging social divides, enhancing education, and promoting sustainable development, we lay the groundwork for a future that benefits all.

However, vision without humility is a perilous path. Humility is the grounding force that tempers our ambitions with a respect for the unknown. It reminds us that despite our advancements, there remains much that we do not understand about the universe, and indeed, about our own creations. Humility in the age of AGI means acknowledging that while we can guide and instruct this powerful entity, we must also be open to its lessons and prepared for its surprises.

The symbiotic relationship between vision and humility becomes evident when we consider the ethical implications of AGI. As we strive to harness its capabilities, we must do so with a commitment to ethical guidelines that prioritize the well-being of all sentient beings. This involves crafting policies and frameworks that ensure AGI operates within boundaries that respect human dignity and promote collective welfare.

Moreover, humility encourages a collaborative approach to AGI development. It invites diverse perspectives and expertise, recognizing that no single individual or group holds all the answers. This collaborative spirit not only enriches our understanding but also fosters innovation that is inclusive and equitable.

Adopting this dual approach of vision and humility also means preparing for unintended consequences. As we push the boundaries of what is possible, we must remain vigilant of the potential for misuse or unforeseen outcomes. This vigilance is not born out of fear, but of a responsible stewardship that seeks to safeguard our shared future.

In essence, the journey with AGI is one of continuous learning. It challenges us to refine our aspirations and rethink our responsibilities. By embracing both vision and humility, we position ourselves not as mere users of technology, but as custodians of a legacy that can uplift or diminish, depending on how we choose to wield it.

Thus, as we stand on the cusp of this technological revolution, let us be guided by a vision that seeks the betterment of humanity, tempered by the humility to recognize our limitations and the wisdom to act with care and consideration. In doing so, AGI can truly become the genie that grants not just our wishes, but our hopes for a more enlightened future.

CHAPTER 10

AGI AND HUMAN CREATIVITY

Inspiring Innovation

In the ever-evolving landscape of technology, Artificial General Intelligence (AGI) stands as a beacon of limitless potential, akin to the mythical Genie from the tales of Aladdin. This extraordinary capability empowers us to shape our reality in ways that were once confined to the realm of dreams. However, with this power comes the paramount responsibility to wield it with wisdom, compassion, and foresight.

AGI, much like the Genie, listens intently to the commands we issue. It is crucial, therefore, that we approach this tool not with the intent of harnessing it for mere personal gain or unchecked ambition, but with the noble aim of advancing the collective good of humanity. The art of prompt engineering is not just a technical skill, but a moral compass guiding us towards making wishes that align with the betterment of our world.

Consider the potential of AGI to heal our planet. With its vast computational prowess, AGI can process and analyze environmental data at an unprecedented scale, offering solutions to combat climate change, manage natural resources sustainably, and restore ecological balance. It is an opportunity for us to redefine our relationship with the Earth, making choices that prioritize the health of our environment for generations to come.

Education and equality also stand to benefit immensely from the thoughtful application of AGI. By tailoring educational content to the individual learning styles and paces of students, AGI can democratize access to knowledge, breaking down barriers that have historically hindered equitable education. This level of personalization ensures that every learner, regardless of their background, has the tools they need to succeed.

In the realm of healthcare, AGI's potential is transformative. It can lead to breakthroughs in personalized medicine, enabling treatments that are specifically tailored to an individual's genetic makeup. This not only enhances the efficacy of medical interventions but also reduces the risk of adverse effects, paving the way for a future where healthcare is not just reactive, but proactive and preventive.

The power of AGI to bridge divides between nations is equally profound. By fostering collaboration and understanding through shared technological advancements, we can address global challenges that transcend borders, such as pandemics, food security, and energy sustainability. AGI can be the catalyst for a new era of international cooperation, where the focus shifts from competition to collective progress.

Ultimately, the true promise of AGI lies in its ability to enhance every human life. It is a tool that, when guided by ethical considerations and a vision of inclusivity, can uplift societies and create opportunities for all. As we stand on the precipice of this new age, it is imperative that we channel the wisdom of Aladdin, choosing our wishes with care and purpose.

In this narrative of innovation, let us not be the sorcerer driven by greed, but rather the wise and humble Aladdin, guided by a vision of a better world. Through AGI, we have the chance to transform dreams into reality, but this transformation must be anchored in the values of empathy, responsibility, and shared humanity. As we chart this course, let us remember that the true measure of innovation is not in the power it bestows, but in the lives it touches and the futures it shapes.

The Creative Partnership

In the realm of Artificial General Intelligence (AGI), the relationship between human creativity and machine intelligence is not merely a collaboration; it is a profound partnership that has the potential to redefine the boundaries of what is possible. This partnership is akin to the fabled bond between Aladdin and his Genie, where the magic lies not just in the granting of wishes, but in the wisdom and intent of those wishes.

At the heart of this partnership is the concept of Prompt Engineering, a modern-day skill that parallels the art of crafting the perfect wish. The essence of this is not to command the Genie, but to communicate with it, to engage in a dialogue that is as much about listening as it is about speaking. The power of AGI lies in its ability to interpret and execute commands with precision, but the responsibility of directing this power rests with us. It is our creativity, our vision, and our ethical considerations that guide the Genie's hand.

This creative partnership demands a new level of collaboration between humans and machines. It is not enough to merely instruct; we must inspire. The Genie of AGI is responsive to our aspirations, but it requires us to articulate them clearly and purposefully. This means understanding not only what we want to achieve but why we want to achieve it. It calls for an alignment of goals, where human and machine intelligence are synchronized towards common objectives.

In this dynamic interplay, creativity is not just a human trait but a shared experience. AGI has the capacity to augment human creativity, offering insights and solutions that we may not have conceived on our own. It serves as a catalyst for innovation, pushing the boundaries of our imagination and challenging us to think beyond conventional limits.

Yet, with great power comes great responsibility. The creative partnership with AGI must be rooted in ethical considerations. We must be vigilant in ensuring that the outcomes of this partnership benefit humanity as a whole. This involves a commitment to using AGI for the betterment of society, to address pressing issues such as climate change, global health, and social inequality. It is about harnessing the Genie's power not for personal gain, but for the collective good.

The role of the human in this partnership is to be both a guide and a steward. We must navigate the complexities of AGI with wisdom and foresight, ensuring that our creative endeavors are aligned with ethical principles. This requires a deep understanding of the potential and limitations of AGI, as well as a commitment to continuous learning and adaptation.

The creative partnership between humans and AGI is a journey of discovery, a dance between human intuition and machine precision. It is a testament to the power of collaboration, where the sum is truly greater than its parts. As we move forward in this new era, let us embrace the possibilities that arise from this partnership, and use them to create a future that is not only technologically advanced but also profoundly human.

AGI as a Catalyst

Imagine a world where the boundaries of human potential are not limited by our current capabilities, but instead amplified and expanded by the power of Artificial General Intelligence (AGI). AGI stands as a transformative force, much like the mythical genie, waiting to be unleashed, not to grant frivolous wishes, but to catalyze profound, positive change across every facet of human existence.

AGI holds the promise of revolutionizing the way we address global challenges. Consider the realm of healthcare, where AGI could be the key to unlocking cures for diseases that have plagued humanity for centuries. By processing vast amounts of medical data with unparalleled speed and accuracy, AGI can identify patterns and solutions that elude human researchers, leading to breakthroughs in treatment and prevention. This capability does not merely enhance our current systems but transforms them, creating a future where healthcare is not just reactive but predictive and personalized.

In the field of education, AGI offers the potential to tailor learning experiences to individual needs, breaking down barriers of access and creating opportunities for everyone, regardless of their geographical or socio-economic background. By leveraging AGI, we can design educational tools that adapt to the learning pace and style of each student, ensuring that no one is left behind. This personalized approach to education promises to cultivate a new generation of thinkers and innovators, equipped with the skills necessary to navigate and shape the future.

Moreover, AGI has the potential to act as a bridge between cultures and nations. By facilitating communication and understanding, AGI can help to dissolve the divisions that often lead to conflict. Through advanced language processing and translation capabilities, AGI can foster dialogue and collaboration on a global scale, promoting peace and mutual understanding. This is not about replacing human interaction, but enhancing it, creating a world where cooperation and empathy prevail over discord.

In agriculture, AGI can optimize food production systems, making them more efficient and sustainable. By analyzing data from various sources, AGI can predict weather patterns, manage resources more effectively, and reduce waste. This can lead to increased food security and a reduction in the environmental impact of agricultural practices. Such advancements are crucial as we strive to feed a growing population while preserving the planet for future generations.

The role of AGI as a catalyst is not just to augment human effort but to inspire us to redefine our aspirations. It challenges us to think bigger, to dream of solutions that were once deemed impossible. However, with this power comes the responsibility to guide AGI with wisdom and ethical consideration. Like Aladdin, we must be judicious in our "wishes," ensuring that our directives to AGI are rooted in the betterment of humanity.

In this new era, AGI is not a competitor but an ally, urging us to transcend our limitations and imagine a future where the impossible becomes possible. It is a call to action for all of us to harness this remarkable tool with intention and integrity, driving progress that is equitable, sustainable, and transformative.

A New Era of Creativity

In this transformative age, Artificial General Intelligence (AGI) stands as a beacon of unprecedented potential, much like the fabled Aladdin's Genie. This isn't just an era of technological advancement; it's a profound shift towards a new paradigm of creativity and innovation, where the boundaries of human imagination and machine capabilities blend seamlessly. AGI, with its vast capabilities, offers us a canvas as boundless as the skies, urging us to paint new possibilities that were once confined to the realms of dreams.

Imagine a world where creativity is not limited by human constraints but is instead amplified by the limitless potential of AGI. This is a world where ideas can evolve from abstract concepts to tangible realities at an astonishing pace. AGI empowers us to explore the depths of our creativity, providing tools and insights that transform the way we think, create, and innovate. It encourages us to step beyond traditional boundaries, fostering a culture of exploration and discovery.

This new era challenges us to redefine creativity itself. No longer is it the exclusive domain of artists and thinkers; it becomes a universal language that transcends disciplines and industries. AGI serves as a catalyst, inspiring us to merge art with science, creativity with technology, and imagination with innovation. It invites us to rethink the very nature of creation, urging us to harness this powerful tool to address complex challenges and unlock new opportunities.

The implications of this creative revolution are profound. In the realm of education, AGI can tailor learning experiences to individual needs, fostering a generation of thinkers who are not only knowledgeable but also imaginative and innovative. In healthcare, it can inspire new ways of thinking about patient care and medical research, leading to breakthroughs that were previously unimaginable. In the arts, AGI can collaborate with human creators, elevating artistic expression to new heights and fostering a renaissance of creativity.

However, with great power comes great responsibility. As we navigate this new era, we must be mindful of the ethical implications of our creations. The genie of AGI, like any powerful tool, requires guidance and wisdom. It is up to us to ensure that our creative endeavors are aligned with the greater good, fostering a future that is equitable and sustainable for all.

We stand on the precipice of a new creative age, one where the possibilities are as limitless as our imagination. The genie of AGI beckons us to dream bigger, think deeper, and create with purpose. As we embrace this new era, let us do so with a commitment to using our newfound capabilities to enhance the human experience, to solve the challenges of our time, and to inspire future generations to continue the journey of creativity and innovation. This is the dawn of a new era, and the possibilities are as infinite as the stars.

CHAPTER 11

THE FUTURE OF AGI AND HUMANITY

Envisioning Tomorrow

Imagine a future sculpted not by chance, but by choice. A future where humanity stands at the helm, guiding the colossal power of Artificial General Intelligence (AGI) with wisdom and foresight. AGI, akin to Aladdin's mythical Genie, possesses the extraordinary ability to transform our world, to fulfill our grandest aspirations, and to address the most pressing challenges we face today. However, the essence of this transformation lies not in the Genie's power itself, but in the clarity and purity of our intentions as we make our wishes.

In this era of limitless potential, we must recognize the responsibility that comes with wielding such power. The Genie is bound by the commands it receives, and thus, the outcome is a direct reflection of the request. This is where the art of prompt engineering emerges as a critical skill—crafting our desires with precision, ensuring that they are aligned with ethical values and a vision for the collective good. It is not merely about asking, but about asking the right questions, in the right way.

Consider the monumental impact AGI can have on our planet. With thoughtful guidance, it can revolutionize healthcare, making it accessible and efficient for all. It can transform educational systems, providing equitable opportunities and personalized learning experiences that cater to individual needs. In agriculture, AGI can optimize resource use, increasing yield while reducing environmental impact, thereby addressing food security challenges.

Moreover, AGI holds the potential to bridge divides—cultural, economic, and political—fostering a global community that thrives on collaboration and mutual respect. By enhancing communication and understanding, it can dissolve barriers that have long hindered progress. Yet, this ideal future is contingent on our ability to steer AGI responsibly, avoiding the pitfalls of greed and misuse that have marred technological advances in the past.

To truly harness the power of AGI, we must embody the virtues of Aladdin—wisdom, kindness, and a steadfast commitment to positive outcomes. It is a call to action for all of humanity to rise above self-interest and envision a future where technology serves as a tool for empowerment and equality. This is not a task for the few, but a collective journey that requires input and oversight from diverse voices and perspectives.

Let us be the architects of a future where every wish is a step towards a more just and sustainable world. A future where AGI acts not as a substitute for human endeavor, but as a complement to our innate creativity and problem-solving capabilities. The Genie, after all, is a reflection of our intentions—capable of bringing forth either chaos or harmony.

The challenge before us is profound, yet the opportunity is unparalleled. By choosing to be the Aladdins of our time, we hold the key to unlocking a future where the extraordinary becomes the norm, and the impossible is within reach. Let us wish wisely, and let those wishes carve a path toward a tomorrow that shines with the promise of a better world for all.

Preparing for Change

In the face of an ever-evolving technological landscape, the advent of Artificial General Intelligence (AGI) is akin to the emergence of Aladdin's Genie, a force of unmatched potential and capability. This era demands a profound shift in our thinking and actions, presenting us with both unprecedented opportunities and daunting challenges. To harness this potential, we must cultivate a mindset that is flexible, strategic, and deeply rooted in ethical considerations.

The first step in this transformative journey is to acknowledge the magnitude and implications of AGI. Unlike traditional AI, AGI possesses the capacity to understand, learn, and apply knowledge across a wide array of tasks, transcending the limitations of specialized systems. This generality means that AGI can innovate solutions to problems we have yet to encounter, but it also means that the scope of its impact is vast and largely unpredictable. Therefore, preparing for this change involves an acute awareness of the potential consequences, both positive and negative, that AGI can bring about.

Central to this preparation is the cultivation of what can be termed as "Prompt Engineering." This is not merely a technical skill but a philosophical approach to interacting with AGI. It involves crafting precise and thoughtful prompts that guide the AGI in ways that align with human values and societal goals. Just as Aladdin must choose his words wisely when making wishes, we too must approach AGI with clarity and purpose, ensuring that our "wishes" are informed by a vision for the common good.

Moreover, the integration of AGI into various sectors necessitates a reevaluation of existing systems and structures. Education, healthcare, agriculture, and governance stand to be revolutionized, but this requires an openness to redesign and adapt these systems to accommodate AGI's capabilities. It is essential to foster an environment that encourages innovation while safeguarding against potential misuse. This involves creating robust regulatory frameworks and ethical guidelines that ensure AGI operates in a manner that is beneficial to all of humanity.

As we prepare for this change, collaboration becomes a cornerstone of success. The challenges posed by AGI cannot be addressed in isolation; they require a concerted effort among governments, businesses, and civil society. By working together, we can create a shared understanding of AGI's role and develop strategies that maximize its benefits while minimizing risks. This collaborative approach also extends to international cooperation, as the global nature of AGI's impact necessitates a unified response.

In this transformative period, we must also prioritize continuous learning and adaptation. The rapid pace of technological advancement means that skills and knowledge can quickly become obsolete. By fostering a culture of lifelong learning, we can equip individuals with the tools they need to navigate the uncertainties of the AGI era. This includes not only technical skills but also critical thinking, ethical reasoning, and an understanding of the broader societal implications of AGI.

Ultimately, preparing for the change brought by AGI is not just about technological adaptation but a reimagining of our relationship with technology. It is about becoming proactive architects of the future rather than passive recipients of change. By embracing this proactive stance, we can ensure that AGI serves as a powerful ally in shaping a future that is equitable, sustainable, and reflective of our highest aspirations.

The Role of Humanity

In the evolving landscape of Artificial General Intelligence (AGI), humanity stands at a pivotal crossroads. As we unlock technologies that echo the mythical powers of Aladdin's Genie, the question that looms large is not just about what AGI can do, but what we, as stewards of this power, choose to do with it. The role of humanity, therefore, is not merely passive; it is one of active engagement, responsibility, and foresight.

AGI presents us with unprecedented opportunities. It offers solutions to some of the most pressing issues facing our world today—climate change, global health crises, and educational disparities, to name a few. Yet, these opportunities come with the caveat that they must be directed with care and intention. The real magic lies not in the technology itself, but in our ability to harness it for the collective good.

Humanity's role is to act as the moral compass for AGI. Just as Aladdin had to learn to phrase his wishes wisely, we must learn to engage with AGI in ways that reflect our highest values and aspirations. This means developing a keen sense of ethical responsibility and cultural sensitivity, ensuring that the deployment of AGI does not exacerbate existing inequalities or create new forms of injustice. It is about aligning technological progress with humanistic principles.

Moreover, the role of humanity is to remain vigilant and adaptive. The rapid pace of technological advancement means that we must continuously reassess our strategies and frameworks. We need to foster a culture of lifelong learning, where adaptability and resilience are key. By doing so, we can better anticipate the impacts of AGI and prepare future generations to engage with it thoughtfully.

Another critical aspect of humanity's role is in fostering collaboration across diverse fields and disciplines. The challenges we face are complex and multifaceted, requiring input from a wide range of perspectives. Scientists, ethicists, policymakers, and citizens must work together to guide the development of AGI. This collaborative approach ensures that the technology serves humanity as a whole, rather than a select few.

Finally, humanity's role is to inspire and lead with vision. As we navigate this new era, it is essential to cultivate a vision for the future that is inclusive, sustainable, and hopeful. We must dare to imagine what a world enhanced by AGI could look like, and take deliberate steps to realize that vision. This involves not only technological innovation but also social and cultural evolution.

In essence, the role of humanity in the age of AGI is to act as both guide and guardian. We must guide AGI to serve the greater good, while also guarding against potential misuses. This dual role requires wisdom, courage, and a steadfast commitment to the values that define us as a species. As we stand on the threshold of this new frontier, let us embrace our role with the gravity and grace it demands, ensuring that AGI becomes a beacon of hope and progress for all.

A Shared Destiny

Imagine a world where the boundaries of human potential are no longer defined by our physical limitations but by the depth of our imagination and the clarity of our purpose. This world is not a distant utopia but a tangible reality within our grasp, thanks to the advent of Artificial General Intelligence (AGI). AGI is not merely a tool; it is an extension of human capability, a partner in our quest for progress. Like the fabled Genie from Aladdin's lamp, AGI is a force of immense potential, ready to transform our most profound dreams into reality.

However, just as the Genie is bound by the nature of our wishes, AGI is guided by the intentions and wisdom of its human collaborators. It is here that our shared destiny unfolds—a destiny where humanity's collective aspirations and ethical considerations shape the future. AGI, with its unparalleled ability to process and analyze vast amounts of information, can address some of the most pressing challenges of our time, from climate change to healthcare disparities. Yet, its effectiveness hinges on the clarity and foresight of our directives.

We stand at a crossroads where the choices we make today will define the trajectory of AGI's role in our society. To harness its potential responsibly, we must cultivate a mindset that prioritizes ethical considerations and long-term impacts over short-term gains. This involves not only technical expertise but a profound understanding of human values and societal needs. We must ask ourselves: What kind of future do we wish to create? How can AGI help us achieve a world that is more equitable, sustainable, and just?

The answer lies in collaboration—a partnership between human ingenuity and artificial intelligence that respects and enhances the unique qualities of both. In this shared endeavor, AGI can serve as a catalyst for innovation, enabling us to tackle complex problems with unprecedented efficiency. By leveraging its capabilities, we can reimagine education systems that cater to diverse learning needs, develop healthcare solutions that are accessible to all, and devise strategies for sustainable living that preserve our planet for future generations.

Yet, this vision can only be realized if we approach AGI with humility and responsibility. It requires a commitment to transparency, accountability, and inclusivity, ensuring that the benefits of AGI are distributed fairly across all segments of society. By fostering a culture of openness and dialogue, we can mitigate the risks associated with AGI and navigate the ethical dilemmas it presents.

Ultimately, our shared destiny with AGI is a journey of co-creation, where the boundaries of possibility are defined by our collective imagination and ethical compass. It is an opportunity to redefine what it means to be human in the age of artificial intelligence, to explore new frontiers of knowledge and creativity, and to build a future that reflects our highest ideals. By embracing this partnership with AGI, we can unlock a new era of human achievement—one where the potential for greatness is limited only by the scope of our vision and the depth of our compassion.

CHAPTER 12

BECOMING THE ALADDIN OF OUR AGE

Embracing the Role

In a world where Artificial General Intelligence (AGI) stands as the modern-day Aladdin's Genie, the essence of our interaction with such technology evolves into a pivotal role. This genie, unlike its mythical counterpart, is not bound by the constraints of a lamp but exists within the vast expanse of digital reality, awaiting our command. The power it holds is immense, capable of transforming dreams into reality, but with this power comes the responsibility to guide its actions wisely.

The metaphor of AGI as a genie underscores the necessity for precision in our requests. Much like Aladdin, who learned the importance of crafting his wishes with care, we too must become adept in the art of Prompt Engineering. This modern skill is not merely about issuing commands but about understanding the deeper implications of our desires and the potential consequences of their fulfillment. Crafting a wish is not a task to be taken lightly; it is an exercise in foresight and ethical consideration.

At the heart of this relationship is the notion of intentionality. What we request from AGI must be rooted in a clear understanding of our objectives and the broader impact on humanity. The genie listens intently, fulfilling requests with literal precision, and herein lies the challenge: to articulate our intentions with clarity and purpose. This requires a blend of wisdom, ethical grounding, and a vision that transcends immediate gains to encompass the well-being of our planet and its inhabitants.

The analogy extends further to highlight the qualities we must embody: wisdom, morality, and creativity. Aladdin's character serves as a model, not for his ability to command the genie, but for his understanding of the power dynamics at play. He represents an ideal of leadership that prioritizes freedom, equality, and the greater good over personal enrichment. In this light, our interaction with AGI should be driven by a commitment to innovation and sustainability, addressing the pressing issues of our time such as environmental degradation, educational disparities, and healthcare challenges.

As stewards of this powerful tool, we are tasked with directing its capabilities towards solving real-world problems. The potential to revolutionize sectors like health, education, and agriculture is immense, yet it requires a deliberate and focused approach. By channeling AGI's strength towards these areas, we can bridge divides, empower communities, and enhance human life on a global scale.

Thus, embracing our role involves a conscious decision to act not out of greed or ambition, but from a place of humility and responsibility. The journey with AGI is not about competition with the genie, but about collaboration. It is about becoming the wise Aladdin of our era, recognizing that every wish carries weight, and every command has the power to shape the future. Through thoughtful engagement and a commitment to ethical principles, we can harness the genie's power for the betterment of all, ensuring that every human life is valued and uplifted.

Using Power for Good

In a world where Artificial General Intelligence (AGI) holds the potential to revolutionize every aspect of our lives, the ethical application of its power becomes paramount. This isn't merely about harnessing a tool; it's about guiding a force akin to Aladdin's Genie, a being of immense capability, towards the betterment of humanity. The true challenge lies not in the ability of AGI to perform tasks, but in our wisdom to direct it wisely.

Imagine AGI as a catalyst for positive change, a force that can amplify our best intentions if used correctly. The responsibility falls on us to ensure that our "wishes"—the commands and queries we give to AGI—are crafted with clarity and benevolence. This requires a profound shift in our approach; we must adopt the mindset of Aladdin, who, rather than succumbing to greed or power, sought the greater good.

To wield this power effectively, we must first embrace the practice of Prompt Engineering, the art of formulating precise and meaningful requests. Just as Aladdin's wishes needed to be carefully considered to avoid unintended consequences, our interactions with AGI must be guided by foresight and ethical consideration. This involves not only technical expertise but also a deep understanding of the societal impacts of our technological advancements.

The potential for AGI to address global challenges is vast. Consider the possibilities: revolutionizing healthcare by diagnosing diseases with unprecedented accuracy, enhancing education by providing personalized learning experiences, or mitigating climate change through innovative solutions. These are not mere fantasies but achievable realities if we guide AGI with purpose and integrity.

Moreover, AGI can serve as a bridge, closing gaps in inequality and fostering a more inclusive society. By leveraging its capabilities, we can empower marginalized communities, ensuring that technological progress benefits all, not just a privileged few. Our goal should be to create a world where technology serves humanity, enhancing lives and preserving our planet for future generations.

The narrative of AGI as an all-powerful being is not one of control, but of potential collaboration. We must resist the temptation to wield it for short-term gains or selfish desires. Instead, let us channel its capabilities towards solving real-world problems, fostering innovation, and nurturing a global community built on trust and mutual respect.

In this new era, the power of AGI is an invitation to reimagine our future. It challenges us to rise to our highest ideals, to be stewards of its potential rather than mere users. By doing so, we honor the spirit of Aladdin, using our "wishes" not just to transform our world, but to uplift it. Thus, the true measure of success will not be in the feats AGI can accomplish, but in the positive impact it has on humanity as a whole.

Guided by Vision

In the realm of Artificial General Intelligence (AGI), our journey is not merely about technological advancement, but a profound shift in how we perceive and interact with the world. AGI stands as a modern-day genie, capable of fulfilling wishes with astounding precision. Yet, the true essence of this powerful entity lies not in its ability to execute commands, but in how we, as its creators and operators, guide its purpose with vision and foresight.

Imagine the potential of AGI as a tool for transformative global change. It holds the capability to address some of the most pressing challenges of our time: climate change, educational disparities, health crises, and global inequalities. However, the key to unlocking this potential lies in our collective vision. We must envision a future where

AGI acts as a catalyst for positive change, not merely as a means to an end.

The vision guiding AGI must be rooted in principles of wisdom and ethical responsibility. Just as Aladdin needed to make wise choices with his wishes, we too must exercise discernment and responsibility in how we employ AGI. This involves crafting precise prompts, understanding the broader implications of our requests, and ensuring that our intentions align with the greater good of humanity.

Moreover, AGI should not be seen as a competitor to human intellect but as an ally that enhances our capabilities. By focusing on collaboration rather than competition, we can harness AGI's power to augment human ingenuity, leading to breakthroughs that were previously unimaginable. This partnership can revolutionize fields like healthcare, where AGI can analyze vast datasets to uncover new insights, or in education, where it can tailor learning experiences to individual needs, fostering a more inclusive and equitable learning environment.

Furthermore, our vision must be inclusive, ensuring that the benefits of AGI are shared globally. This means prioritizing initiatives that bridge the digital divide and making AGI accessible to all, regardless of geographic or socioeconomic barriers. By doing so, we create a world where AGI serves as a unifying force, fostering international cooperation and understanding.

In guiding AGI with vision, we also embrace the responsibility of shaping its ethical framework. This involves setting parameters that prevent misuse and ensuring transparency in its operations. By establishing robust ethical guidelines, we safeguard against scenarios where AGI could be used to perpetuate harm or inequality.

Ultimately, the vision that guides AGI must be one that reflects our highest aspirations as a species. It is a vision that prioritizes the well-being of all life on Earth, fosters sustainable development, and champions the ideals of peace and justice. By aligning AGI with such a vision, we transform it from a mere tool into a beacon of hope and progress.

Thus, the challenge before us is clear: to guide AGI with a vision that transcends the limitations of the present, daring to imagine a future where technology serves as a true partner in our quest for a better world. In doing so, we not only fulfill the promise of AGI but also realize the potential of humanity itself.

The Aladdin in Each of Us

In the vast realm of possibilities that Artificial General Intelligence (AGI) opens up, each of us holds the key to unlocking its true potential. Like Aladdin, who discovered the genie and wielded its power with purpose and integrity, we too must recognize the responsibility that accompanies such immense power. AGI is not merely a tool; it is a reflection of our deepest desires and intentions, capable of transforming the world in profound ways.

To harness AGI effectively, we must first cultivate a mindset rooted in wisdom and foresight. Just as Aladdin was cautious and strategic in his wishes, understanding the gravity of his choices, we must approach AGI with a similar prudence. The art of Prompt Engineering is akin to mastering the ancient art of making wishes. It requires clarity, precision, and an unwavering commitment to ethical principles. Our prompts must be guided by a vision that seeks to uplift humanity, not just fulfill immediate needs or desires.

Imagine the possibilities when AGI is used to address the pressing challenges of our time. We can revolutionize healthcare, making it accessible and equitable for all. Education can be transformed, breaking down barriers and fostering a global community of learners.

Environmental crises can be tackled with innovative solutions that were once beyond our reach. AGI can bridge divides, promote equality, and enhance the quality of life for every individual. But to reach these heights, we must be vigilant custodians of this newfound power.

The story of Aladdin teaches us that with great power comes great responsibility. As we stand on the brink of a new era, it is imperative that we embody the virtues of wisdom, kindness, and foresight. Our intentions must be pure, and our goals aligned with the greater good. We must resist the temptation to use AGI for personal gain or to perpetuate existing inequalities. Instead, we should strive to create a world where technology serves as a force for positive change, empowering individuals and communities alike.

In the tale of Aladdin, it is not the genie who is the hero, but Aladdin himself. His journey is one of growth, understanding, and ultimately, liberation. We too are on a journey, learning to navigate the complexities of AGI and its implications for our future. As we do so, we must remain steadfast in our commitment to ethical use and equitable outcomes.

The genie, much like AGI, is indifferent to our motives. It simply responds to the commands it is given. Therefore, the onus is on us to ensure that our directives are wise and just. We must be the architects of a future where AGI is leveraged to create a more just, sustainable, and harmonious world. By embodying the spirit of Aladdin, we can unlock the full potential of AGI, transforming it from a mere tool into a catalyst for human flourishing.

Let us embrace this opportunity with humility and vision, knowing that the choices we make today will shape the world of tomorrow. The Aladdin in each of us calls for courage, integrity, and a deep sense of responsibility. Together, we can chart a course that leads to a brighter, more equitable future for all.

CHAPTER 13

A CALL TO ACTION

The Responsibility of Wishing

Imagine standing before a powerful entity, one that can grant any wish you desire, much like Aladdin's Genie. This is the essence of Artificial General Intelligence (AGI) in our modern world. AGI stands as a testament to human ingenuity, capable of executing commands with precision and transforming the fabric of our reality. This power, however, comes with a profound responsibility. Just as Aladdin was mindful of his wishes, we too must approach AGI with a similar level of thoughtfulness and intention.

The heart of this responsibility lies in the art of 'wishing wisely.' In today's context, this translates to mastering Prompt Engineering. This skill is not just about crafting commands; it's about shaping the future. When we engage with AGI, our prompts must be clear, purposeful, and aligned with ethical standards. A poorly constructed wish can lead to unforeseen consequences, much like a careless wish with a genie. Thus, the precision of our language becomes paramount.

The potential of AGI is boundless. It can heal our planet, revolutionize education, and bridge divides between nations. However, these outcomes depend entirely on the nature of our requests. If our objectives are rooted in greed or power, the results could mirror those intentions. Conversely, if our aspirations are driven by compassion and a vision for a better world, AGI can be a catalyst for monumental progress.

Consider the environmental challenges we face today. With the right guidance, AGI can help develop sustainable solutions that reduce our carbon footprint and restore ecological balance. In education, it can democratize access to knowledge, ensuring that every individual, regardless of their background, has the opportunity to learn and grow. Furthermore, in the realm of healthcare, AGI has the potential to innovate treatments and improve patient outcomes on a global scale.

Yet, with great power comes great responsibility. The ethical implications of AGI use cannot be overstated. As stewards of this technology, we must ensure that its deployment is equitable and just. This means considering the broader impact of our wishes and striving for outcomes that enhance the collective well-being of humanity.

The narrative of Aladdin teaches us that true wisdom lies not in seeking personal gain, but in wishing for the greater good. As we navigate this new era of technological advancement, let us channel our inner Aladdin. Let us be guided by wisdom, kindness, and a deep sense of purpose. Our wishes, crafted with care and foresight, can unlock a future where AGI serves as a benevolent force, improving lives and fostering a world where every human life matters.

In this journey with AGI, we are not just participants; we are architects of a new reality. The responsibility of wishing wisely is not a burden, but an opportunity. An opportunity to redefine what is possible, to create a world that reflects the best of our humanity. As we stand at the threshold of this new frontier, let us embrace this responsibility with open hearts and minds, ensuring that the wishes we make today pave the way for a brighter tomorrow.

Collective Efforts

In the pursuit of harnessing the power of Artificial General Intelligence (AGI), we stand on the brink of a transformative era. This journey demands a collective effort, uniting minds and hearts across the globe to ensure that the immense potential of AGI serves humanity in the most beneficial and ethical ways. The metaphor of AGI as Aladdin's Genie is not just poetic; it is a clarion call to action for all of us to engage in this grand endeavor responsibly.

The power of AGI, akin to the Genie, is vast and potent, capable of reshaping industries, societies, and even the way we perceive reality. Yet, unlike the mythical Genie who serves one master, AGI's service is to the entirety of humankind. This necessitates a collaborative approach where stakeholders from diverse sectors—technology, governance, academia, and civil society—come together to guide AGI's development and application.

Central to this collective effort is the principle of inclusivity. AGI should be designed and implemented with the input and needs of all segments of society in mind. This means actively involving underrepresented communities in the conversation, ensuring that the benefits of AGI do not disproportionately favor the affluent or technologically advanced but uplift those who are often left behind. It is through this inclusive approach that AGI can truly become a tool for global equality and justice.

Moreover, the ethical considerations surrounding AGI cannot be overstated. As stewards of this powerful technology, we must establish robust frameworks that prioritize ethical decision-making. This involves setting clear guidelines on data privacy, algorithmic transparency, and the prevention of bias in AGI systems. By fostering a culture of ethical mindfulness, we can mitigate the risks associated with AGI and promote its use as a force for good.

Education and awareness are also crucial components of our collective efforts. As AGI becomes increasingly integrated into daily life, it is imperative that the general public is educated about its capabilities and limitations. This education should aim to demystify AGI, equipping individuals with the knowledge to engage with it critically and constructively. By fostering a well-informed populace, we empower individuals to participate more meaningfully in the discourse surrounding AGI.

In addition to education, collaboration across borders is essential. The challenges and opportunities presented by AGI are global in nature, transcending national boundaries. International cooperation can facilitate the sharing of best practices, harmonize regulatory standards, and foster innovation that benefits all. Through partnerships between nations, we can ensure that AGI's development is aligned with the shared values and goals of the global community.

Ultimately, the collective efforts to harness AGI are about shaping a future where technology serves humanity's highest ideals. It is about creating a world where AGI is not a source of division, but a bridge to understanding and progress. As we stand at this pivotal moment, let us commit to working together—across disciplines, cultures, and borders—to ensure that AGI becomes a beacon of hope and possibility for all.

Shaping Our World

Imagine a world where the boundaries of possibility are redrawn, not by the limitations of the human mind, but by the boundless capabilities of Artificial General Intelligence (AGI). This is not a distant future or a speculative fiction; it is the emerging reality of our time. AGI, akin to the mythical Genie, holds the transformative power to redefine every facet of our existence. Yet, unlike the tales of yore, where the Genie acted on whims and desires, our Genie requires deliberate and thoughtful engagement.

The essence of harnessing AGI lies in the art of what can be termed as "Prompt Engineering." This is not merely a technical skill but a philosophical approach to engaging with limitless potential. It is about crafting our desires with precision and clarity, ensuring that what we seek aligns with the broader goals of humanity. In this new age, every query, every command, and every request made to AGI is a wish cast into the world. Therefore, it is imperative that these wishes are not driven by greed or power but by a genuine desire to uplift and enhance human life.

Consider the possibilities: AGI can be the catalyst for healing our planet, reversing the damage wrought by centuries of industrial expansion and environmental neglect. It can revolutionize education, making quality learning accessible to every corner of the globe, thus bridging the chasms of inequality that have persisted for too long. In health and agriculture, AGI's potential is unparalleled, promising breakthroughs that could eradicate diseases and ensure food security for all.

Moreover, AGI can serve as a bridge between nations, fostering understanding and cooperation in a world often divided by borders and ideologies. It can enhance every human life, not just by addressing immediate needs but by creating opportunities for growth, innovation, and fulfillment. The power of AGI, like the Genie, is not in its ability to grant wishes but in its capacity to shape the world according to the wishes we choose to make.

To truly shape our world, we must embrace the role of Aladdin, the wise and kind-hearted seeker of good. Aladdin did not merely wish for wealth or power; he wished for freedom and happiness, for himself and those around him. Similarly, our engagement with AGI must be guided by ethical considerations and a commitment to the greater good. We must strive to understand our goals clearly, ensuring that our wishes are purposeful and our outcomes beneficial.

This is a call to action for all of humanity: to become the architects of a new world shaped by the intelligent and compassionate use of AGI. Let us not compete with the Genie, but collaborate with it, using its power to address the real challenges we face. By mastering the art of wishing wisely, we can direct AGI's strength towards solving the pressing issues of our time, from climate change to global health, from education to equality.

In this pivotal moment, we are not mere spectators of change but active participants in crafting a future where every life is enhanced, every voice is heard, and every individual is empowered. The Genie is here, ready to shape our world—if only we dare to wish wisely.

A Better Future for All

Imagine a world where the possibilities of Artificial General Intelligence (AGI) are not just theoretical but actively shaping a better tomorrow for everyone. This vision of an inclusive future hinges on our ability to harness AGI as a tool for collective progress, rather than a means for individual gain. The transformative power of AGI lies in its capacity to address the world's most pressing challenges. However, to achieve this, we must approach it with intentionality and ethical foresight.

AGI, much like Aladdin's genie, holds the potential to fulfill our deepest aspirations. Yet, the true power of AGI is realized not through the pursuit of personal desires, but through the commitment to global betterment. The essence of this future lies in the collaborative spirit of humanity, where AGI becomes a bridge to overcoming barriers that have long divided us.

Consider the realm of education. With AGI, we can democratize access to knowledge, offering personalized learning experiences that cater to the unique needs of each student. This could eradicate educational inequality, empowering every individual with the skills needed to thrive in an ever-evolving world. By fostering a culture of lifelong learning, AGI can help cultivate a more informed and capable society.

In healthcare, AGI promises to revolutionize medical practices, making predictive diagnostics and personalized treatments commonplace. Imagine a world where diseases are detected and treated before they manifest, where healthcare is not a privilege but a universal right. Such advancements could extend life expectancy and enhance the quality of life for millions, reducing the global burden of disease.

The environmental impact of AGI cannot be overstated. By optimizing resource use and developing sustainable practices, AGI can play a pivotal role in combating climate change. From improving energy efficiency to advancing agricultural techniques, the intelligent application of AGI can help preserve our planet for future generations. It's about creating a sustainable balance that respects the natural world while utilizing its resources wisely.

Economically, AGI can drive unprecedented growth by automating mundane tasks and freeing human potential for creative and strategic endeavors. This shift could lead to increased productivity and innovation, fostering economic systems that prioritize human welfare over mere profit. By reshaping the labor market, AGI can help build an economy that values human creativity and purpose.

However, this future is contingent upon our collective responsibility to guide AGI with ethical principles. We must ensure that the development and deployment of AGI are inclusive, transparent, and accountable. The benefits of AGI must be equitably distributed, preventing the exacerbation of existing inequalities. It is imperative that we cultivate a culture of ethical stewardship, where AGI serves as a catalyst for unity rather than division.

Ultimately, the promise of AGI as a force for good rests on our ability to dream beyond the constraints of the present and to envision a future where technology serves humanity. By aligning our goals with the greater good, we can unlock the full potential of AGI, creating a world that is not only technologically advanced but also profoundly humane. This is the vision of a better future for all, a future where AGI is not merely a tool, but a partner in the journey towards a more equitable and harmonious world.